Architecture and Space Re-imagined

As with so many facets of contemporary Western life, architecture and space are often experienced and understood as a commodity or product. The premise of this book is to offer alternatives to the practices and values of such Westernised space and Architecture (with a capital A), by exploring the participatory and grass-roots practices used in alternative development models in the Global South. This process re-contextualises the spaces, values, and relationships produced by such alternative methods of development and social agency. It asks whether such spatial practices provide concrete realisations of some key concepts of Western spatial theory, questioning whether we might challenge the space and architectures of capitalist development by learning from the places and practices of others.

Exploring these themes offers a critical examination of alternative development practices methods in the Global South, re-contextualising them as architectural engagements with socio-political space. The comparison of such interdisciplinary contexts and discourses reveals political, social, and economic resonances inherent between previously unconnected spatial protagonists. The interdependence of spatial issues of choice, value, and identity are revealed through a comparative study of the discourses of Henri Lefebvre, John Turner, Doreen Massey, and Nabeel Hamdi. These key protagonists offer a critical framework of discourses from which further connections to socio-spatial discourses and concepts are made, including post-marxist theory, orientalism, post-structural pluralism, development anthropology, post-colonial theory, hybridity, difference, and subalterneity.

By looking to the spaces and practices of alternative development in the Global South this book offers a critical reflection upon the working practices of Westernised architecture and other spatial and political practices. In exploring the methodologies, implications, and values of such participatory development practices this book ultimately seeks to articulate the positive potential and political of learning from the difference, multiplicity, and otherness of development practice in order to re-imagine architecture and space.

Richard Bower is a Lecturer in Architecture and Critical Spatial Theory at The University of Central Lancashire, UK.

Routledge Research in Place, Space and Politics

Series Edited by Professor Clive Barnett
Professor of Geography and Social Theory, University of Exeter, UK.

This series offers a forum for original and innovative research that explores the changing geographies of political life. The series engages with a series of key debates about innovative political forms and addresses key concepts of political analysis such as scale, territory and public space. It brings into focus emerging interdisciplinary conversations about the spaces through which power is exercised, legitimized and contested. Titles within the series range from empirical investigations to theoretical engagements and authors comprise of scholars working in overlapping fields including political geography, political theory, development studies, political sociology, international relations and urban politics.

For a full list of titles in this series, please visit www.routledge.com/series/PSP

Architecture and Space Re-imagined

Learning from the difference, multiplicity, and otherness of development practice

Richard Bower

Routledge
Taylor & Francis Group

LONDON AND NEW YORK

First published 2017
by Routledge

2 Park Square, Milton Park, Abingdon, Oxfordshire OX14 4RN
52 Vanderbilt Avenue, New York, NY 10017

Routledge is an imprint of the Taylor & Francis Group, an informa business

First issued in paperback 2020

British Library Cataloguing in Publication Data
A catalogue record for this book is available from the British Library

Library of Congress Cataloging in Publication Data
Names: Bower, Richard, 1984- author.
Title: Architecture and space re-imagined : learning from the difference,
multiplicity, and otherness of development practice / Richard Bower.
Description: New York : Routledge, 2016. |
Series: Routledge Research in Place, Space and Politics |
Includes bibliographical references and index.
Identifiers: LCCN 2016004399 | ISBN 9781138934146 (hardback) |
ISBN 9781315678146 (e-book)
Subjects: LCSH: Space (Architecture)--Social aspects--Southern
Hemisphere. | Space (Architecture)--Political aspects--Southern Hemisphere.
Classification: LCC NA2765 .B69 2016 | DDC 720/.1--dc23
LC record available at http://lccn.loc.gov/2016004399

ISBN: 978-1-138-93414-6 (hbk)
ISBN: 978-0-367-66829-7 (pbk)

Typeset in Times New Roman
by Taylor & Francis Books

Contents

Preface

It is hard to define a single point of origin for this book. Perhaps this inexact origin is reflected positively in the explorative trajectory of the work,[1] but in essence the questions underlying this research emerged from experiences, discussions, questions, and moments of critical self-reflection that have arisen over the course my architectural education and work in practice. Initially these questions emerged in direct relation to my personal experiences, relationships, and context which I believed were a unique perspective on the challenges and problems of architectural practice. Yet these personal questions reflect much the same often unspoken melancholy, self-doubt, and frustration that afflict many other architectural practitioners and students who struggle with the reality of contemporary practice and life.

Whilst learning and working, researching, and eventually teaching within the discipline of architecture I have repeatedly found myself confronted and frustrated by the implications of architecture's troubled relationship with questions of space, society, and people. This constellation of interconnected 'big(ger) issues' seems to define architecture from the outside inwards, leaving the profession with a self-image and public identity that appears largely unaware, unconcerned, or disengaged with its relationality to the wider world.[2]

Frustratingly, whilst I was training to be an architect I realised that I did not want to be an architect. That is to say, I did not want to spend my life working in the patterns and processes that seemed to define the 'real-world' of (what I can now retrospectively define as) commercial boom-and-bust neoliberal economics, and the demonstrably abstract and self-referential architectural formalism and aesthetics.[3] Whilst that may sound melodramatic and self-serving, I believe (and anecdotally know) that I am by no means unique in feeling trapped by the assumptions and inevitability of Westernised space and architectural practice. Like many other architectural graduates in 2008 I was faced with the reality of trying to work in a commercial architecture context defined by the 'credit-crunch' financial crisis and the oncoming politics of conservative economic austerity. Whilst in retrospect this was not an entirely unique socio-economic moment in history, it was the first 'financial crisis' that I had experienced, and the reality of the situation helped to concretise the gap that had formed between myself and architecture. I found (and in many ways still find) myself locked within a concept of architectural space and identity that I cannot reconcile with my own critical engagement with the political, social, and economical space that defines conventional architecture assumptions of inevitability.

Ultimately the question of how to escape the socio-economic prescriptions and limitations of conventional architecture remains perhaps one of the most prevalent and pressing questions facing the concept and profession of the architect in Westernised space and culture.[4] Thus, whilst this condition of apathy and abstraction is by no means limited to architecture, its processes and relationships are a valuable cipher through which to perceive wider socio-political issues. Similarly, through my own personal journey and experiences with architecture it has become the critical lens by which I have learnt to interpret the wider world around me. In this book I have sought to use architecture and spatial practices as a critical lens with which to engage with the far broader questions of identity and values that pervade the capitalist and neoliberal Westernised space of the Global North.

Acknowledgement

I owe an impossible debt to my partner Nicole who has been with me throughout the twelve years that culminate in the completion of this book. She is the only person who knows the true cost of pursuing this research. I owe her more than any length or combination of mere words can possibly convey. Suffice to say without her this would not have been possible, nor would I be the person I am today.

Notes

1 It is increasingly clear that my own relationship with space and architecture is intrinsically connected to the critical observations drawn in this book, and that my own story is the underlying point of origin for the questions themselves.
2 This concept of '*relationality*' is intrinsic to later discussion of Doreen Massey's articulations of spatial relationships. For now, it is sufficient to simply describe relationality as a conception of relationships that is dynamic and fluid rather than fixed and structural.
3 This was despite being very fortunate in the practices that I did work for. Despite coming very close, by chance I have never been forced by circumstances to work with the type of lowest-common-denominator commercial practice that afflicts the lives, experiences, and aspirations of many architectural graduates. I worked on commercial developments, but reasonably good ones. And I was lucky enough to always seem to end up working with genuinely nice people. The allure of a simple architectural existence, regular employment, money, and security was indeed powerful. To this day if you catch me on the wrong day I will still question whether I should have just become an architect and had a relatively easy and simple life. Humility aside, I was considered by my peers to be good at what I did. And since leaving my last full time work in practice (at time of writing) I am yet to have a single year of economic or personal stability, instead spending my time pursuing the academic adventures that have lead to this book. If there are other individuals out there in

the same boat, then I hope this note means something to you, and as this book unfolds I hope you will see signs of hope and positivity hidden in this work that help to drive you onwards.

4 This fractious relationship between architectural students and the apparent reality of 'real' architecture defines much contemporary debate between architectural educators, academics, and practitioners. Indeed, academics and practitioners have confronted these issues for certainly the past century since the height of the Beaux-Arts model of architectural teaching, through ideological and indoctrinated structuralist modernism, and on to the (relatively contemporary) emergence of an interdisciplinary academic criticality towards the symbolic power of architecture within space. This is not the place to discuss such a complex issue, though I hope that implicit elements within this book may help to spark and re-orientate the frustratingly introspective and unimaginative debate that continues to surround the relationship of architectural pedagogy and practice – the identity and the profession.

Introduction

The premise of reimagining architecture and space (ex)poses an inordinate range of questions, presumptions, and challenges that are undoubtedly impossible to explore in a single book. However, it is important to hold onto the aspiration and spirit inherent within the title and its critical premise: are there other perspectives from which to reimagine architecture and space? What can academics and practitioners learn from examples of alternative models of development, from participatory spatial practices, and from unexpected places and different realities? Might such an alternative reimagining of space reveal opportunities to change the way architecture is produced? Might it also offer the chance to question and challenge the social relations that produce our Westernised space and lives?

In response to these (admittedly huge) questions, this book explores and compares socio-spatial methodologies of development and spatial practices that have been utilised in contexts of pronounced inequality, difference, multiplicity, and otherness. The explicit intent of introducing such examples is to challenge the conventionally held assumption that Westernised space and architecture – driven by neoliberal economics and plutocratic capitalist politics – is an inevitability, and that the social relationships that produce such space and architecture cannot be contested, reimagined, and changed.

However, whilst Westernised space and architecture is implicitly held under scrutiny throughout this book, it is kept largely absent from the critical analysis laid out in the principle chapters. This absence avoids producing another rendition of the all-to-familiar academic critical and negative analysis of Westernised architectural practice, whilst also providing an opportunity to pursue an alternative trajectory of enquiry. Thus in essence, this book seeks to connect the critical spatial advocacy of Western academic theory to alternative examples of participatory spatial practice from the Global South.

In this analysis such methods of alternative practices are thus re-valued both for their ability to produce spaces of architectural humility, social sustainability, and political agency, but also for the tacit and reflexive critique of Westernised space and neoliberal architectural practice that is inherent in their alternative reimagining spatial agency and practice.

A Framework of Critical Trajectories

In attempting to explore a novel comparison of Western theory to alternative practice this book intentionally strays from the conventional path of architectural discourse and it's conventionally self-referential or abstract methods of critique.

In contrast to the assumptions and conventions of Westernised architecture that are produced by neoliberal economics, this book explores alternative methodologies and simple practical realisations of positive socio-spatial agency. These examples are drawn from development practices in informal settlements and contexts of economic absence, most notably in the Global South. A methodology of close textual reading and critical comparison reframes such examples as practical realisations of key Western socio-spatial theories that have defined academic responses to post-Second World-War capitalism.

This premise is extremely open and ambitious, yet these research aspirations are framed around simple critical comparisons between a primary framework of four key protagonists: John F.C. Turner and Nabeel Hamdi as development practitioners, and Henri Lefebvre and Doreen Massey as Western spatial theorists. The relationships between each of these four spatial advocates provide the opportunity to explore their individual intra-disciplinary concepts and achievements, but more provocatively, the interdisciplinary comparisons between distinctly different engagements with space. Thus, in attempting to question the assumptions of Westernised space and architecture we will confront and contest seminal disjunctions and dialectics that remain at the core of academic debate: theoretical and practical, centre and periphery, formal and informal, us and them, self and other, Global North and Global South.

In seeking to critically explore such a broad yet interconnected range of contexts, concepts, and contradictions it is explicitly recognised here and throughout the book that these comparisons cannot (and do not seek to) provide a precise and fixed resolution of any of these interconnected questions, comparisons, and ideas. Instead the premise is to problematise connections and comparisons between discourses in order to first reveal new ways to re-read and utilise existing ideas, and then subsequently hopefully to spark new discussions of the invaluable yet overlooked intersection of theory and practice from the Global North and Global South.

The relationship between architecture, social space, and participation has been repeatedly contested by Blundell Jones et al (2005), Grieg Crysler (2003), Hughes and Sadler (1999), Miessen and Basar (2006), and Jemery Till (2009, pp. 7–12). Similarly, documentations of alternative and participatory architectural agency can be recognised in the discourses of Hassan Fathy (1976), Bernard Rudolfksy (1987), and Giancarlo de Carlo (1949). However, within the premise of this research is an observation that development practice has generally remained side-lined and abstracted to the periphery of architectural discourse. The momentary fascination and subsequent peripheralisation of alternative and participatory architectures is best exemplified by John

Turner's special edition of the Architecture Design journal (1963) which remained for decades the high point of architectural engagement with participatory practices and the alternative spatial production of the Global South.

More recently alternative forms, models, and practices of architecture have emerged within contemporary architectural practice and have been critically contested in academic discussion (Dovey and King, 2013; Sanyal et al., 2008). Perhaps most widely observed are the positive contemporary practices of Teddy Cruz at the US Mexico border (Cruz, 2012, 2011, 2005), the work of Urban Think Tank (UTT) in Caracas and South Africa (Brillembourg and Klumpner, 2012), and the work of Elemental architecture in Chile (Aravena and Lacobelli, 2012).

Yet these various examples have been largely examined from explicitly practical perspectives and in fragmented discourses that have yet to coalesce into a clear theoretical discourse, nor articulate the interconnections of wider spatial discourse to such socially innovative practices. Whilst attempts have been made at such theoretical comparisons of participatory architecture they continue to be in imminent danger of remaining isolated as objects of peripheral intrigue, instead of being contested as viable practices in direct opposition to conventional Westernised architecture (Dovey, 2013; Fraser, 2012; Hughes and Sadler, 1999). It is in the context of this disjunctive gap between participatory development practices and Western theory that this book proposes to contest and interrogate the social, political, and economic value and implications of informal space and alternative methodologies of spatial practice.

In response to presumptions of the inevitability of development towards Western capitalistic hegemony (Massey, 1999, p. 64), our premise is thus to explore the positive theoretical implications of alternative spatial practices of the Global South in order to implicitly speculate on the reflective potential for their appropriation to the Global North. The critical comparisons in the coming chapters seek to demonstrate that in development practice methodologies and informal settlements we can observe spatial, economic, and social relations that offer realisations of Western theoretical aspirations for politicised space.

Here, it is important to explicitly note that this book does not seek to overlook the indescribable inequality and criminal human indecency that distinguishes between everyday life in the Global North and Global South, or the same chasm of inequality that increasingly exists within almost touching distance in cities across the world. This book does not seek to glorify or fetishize the torturous poverty experienced by a vast proportion of the world's population, and openly admits the opportunity to even write this book is based upon my own luck to have been born in a country built upon the ravages of past-empires.

Throughout this pursuit of a critical reimaging of space and architecture there remains an underlying narrative that is both explicitly and implicitly referenced throughout the book in a variety of contexts and theoretical guises. This is the question of value(s). The question of the social, economic, and political values that architecture defined by, and more importantly, whose values are they? This underlying question of value provides constant point of reference throughout an explorative interdisciplinary contestation of spatial practice and theory, and acts as a narrative strand that weaves it way throughout the book.

Thus, in spite of the socio-economic absence and extreme inequality that is inextricably linked with informal space there are elements of social, spatial, and political agency that have emerged from the natural richness and human ingenuity of people that deserve to be recognised beyond mere aesthetics and poverty porn. This book proposes that aspects of informal space and development practice reflect the everyday realisation of key ideas that underpin critical spatial theory. It will suggest that informal space offers qualities of social and political agency that are now almost entirely missing within the neoliberal capitalist contexts and conventional Westernised space and architecture.

Structure

The five main chapters of this book each explore comparisons between two of the main four protagonists. Across these explorations of the four key protagonists' interconnected discourses a series of thematic connections, comparisons, and resonances emerge – authority, choice, freedom, difference, identity, dialogue, value. These themes and connections overlap in-between the respective individual chapters, generating a trajectory of critical analysis that interconnects throughout the whole book before culminating in a concluding chapter that seeks to synthesise a provocative reimagining of space and architecture.

Chapter one introduces and contextualises a premise that the development practice of Turner can be compared to Lefebvre's spatial discourse. This comparison generates a critical lens through which to reveal and interrogate Turner's practical approaches of user-defined housing and self-help development, and Lefebvre's spatial articulations of dialectical materialism and autogestion.

Building on this comparative methodology, chapter two connects and compares Lefebvre's legacy and theoretical lineage concerning cities and space through to the contemporary spatial theory of Doreen Massey. In contrast to previous examination of Massey's spatial interpretation of Marxism emerging only from Althusser (Featherstone and Painter, 2013, p. 4), this chapter provides an alternative comparison founded on the observation that both Lefebvre and Massey propose positive political potential of the social relations of space as a medium for difference and multiplicity.

Chapter three offers renewed comparisons and confrontations of critiques raised in the previous two chapter trajectories, contrasting notions of participation and hierarchy, authority and choice, practice and product. Building on Massey's analysis of Chantal Mouffe and Ernesto Laclau, the concepts of hegemony and geometries of power are introduced in comparison with the contemporary development practices of Nabeel Hamdi. This renewed interdisciplinary intersection contests Hamdi as an exemplar of alternative and positive methodologies, utilising practices of disruption, catalysis, and small socio-spatial changes to deliver sustainable and scaleable social space.

Pursuing the questions of authority, identity, and practice raised in chapter three, chapter four offers a comparison and contextualisation of the historical and critical trajectory of development practice from Turner to Hamdi. This analysis engages with important discourses and historical influences on the evolution of development practice in comparison with theoretical discussions of postcolonial identity and values. Subsequently, this critical comparison observes the methodological evolution of development practice from Turner to Hamdi as mirroring several key notions from postcolonial and post-development theory.

Finally, building upon this post-structural analysis of development practice, chapter five pursues various further contemporary interdisciplinary comparisons of Hamdi's methodologies of practice as contestations of postcolonial identity and values. This analysis reveals theoretical and practical connections of his socio-spatial and economic methodologies to the work of post-modern anthropology as well as the cultural theory of Homi K. Bhabha and Gayatri Spivak. Here these critical comparisons appropriate and leverage ideas drawn from the post-modern anthropological advocacy for ethnographic spatial praxis of coevalness, mutuality, and equality when interacting with the multiplicity and difference of *other* communities.

Chapter six concludes with a synthesis of the overall ideas explored in the book before seeking to reframe our original premise: Can Westernised architecture and space be reimagined by engaging with concepts of difference, multiplicity, and otherness? What might such a world be like? And (how) can we ever get there?

References

Aravena, A., Lacobelli, A., 2012. *Alejandro Aravena: Elemental: Incremental Housing and Participatory Design Manual.* Hatje Cantz, Ostfildern, Germany.

Blundell Jones, P., Petrescu, D., Till, J. (Eds), 2005. *Architecture and Participation.* Spon Press, London.

Brillembourg, A., Klumpner, H. (Eds), 2012. *Torre David: Anarcho Vertical Communities.* Lars Müller, Zurich.

Cruz, T., 2005. Tijuana Case Study Tactics of Invasion: Manufactured Sites. *Architectural Design 75*, 32–37. doi:10.1002/ad.133

Cruz, T., 2011. Latin America Meander: In Search of a New Civic Imagination. *Architectural Design 81*, 110–118. doi:10.1002/ad.1248

Cruz, T., 2012. Mapping Non-Conformity: Post-Bubble Urban Strategies. Hemispheric Institute E-Misférica. Available at http://hemi.nyu.edu/hemi/en/e-misferica-71/cruz (accessed 29 February 2016).

Crysler, C.G., 2003. *Writing Spaces: Discourses of Architecture, Urbanism and the Built Environment, 1960–2000.* Routledge, London.

de Carlo, G., 1949. The Housing Problem in Italy. *Freedom.*

Dovey, K., 2013. Informalising Architecture: The Challenge of Informal Settlements. *Architectural Design 83*, 82–89.

Dovey, K., King, R., 2013. Interstitial Metamorphoses: Informal Urbanism and the Tourist Gaze. *Environment and Planning D: Society and Space 31*, 1022–1040.

Fathy, H., 1976. *Architecture for the Poor: An Experiment in Rural Egypt.* University of Chicago Press, Chicago, IL.

Featherstone, D., Painter, J., 2013. There is no Point of Departure: The Many Trajectories of Doreen Massey, in: Featherstone, D., Painter, J. (Eds), *Spatial Politics: Essays for Doreen Massey.* Wiley-Blackwell, Chichester.

Fraser, M., 2012. The Future is Unwritten: Global Culture, Identity and Economy. *Architectural Design 82*, 60–65.

Hughes, J., Sadler, S., 1999. *Non-Plan: Essays on Freedom, Participation and Change in Modern Architecture and Urbanism.* Routledge, London.

Massey, D., 1999. *Power-Geometries and the Politics of Space-Time.* Department of Geography, University of Heidelberg, Heidelberg.

Miessen, M., Basar, S., 2006. *Did Someone Say Participate? An Atlas of Spatial Practice.* MIT Press, Cambridge, MA.

Rudolfsky, B., 1987. *Architecture without Architects: A Short Introduction to Non-Pedigreed Architecture,* Reprinted Edition. University of New Mexico Press, Albuquerque.

Sanyal, B., Rosan, C., Vale, L.J. (Eds), 2008. *Planning Ideas That Matter: Livability, Territoriality, Governance and Reflective Practice.* MIT Press, Cambridge, MA.

Till, J., 2009. *Architecture Depends.* MIT Press, Cambridge, MA.

Turner, J.F.C., 1963. Dwelling Resources in South America. *Architectural Design 8*, 360–393.

1 Dialectical materialism and participatory housing

In this first chapter we will explore and contextualise the premise that the anarchist development practice of John F.C. Turner can be compared to the (post)Marxist theory of Henri Lefebvre. However, the thematic observations of this comparison first require the explication and appropriation of various theoretical concepts that will form the basis for the trajectories of more exploratory chapters later.

At first glance Turner and Lefebvre are perhaps an unlikely pairing to discuss. Their works have each defined paradigmatic shifts in their respective fields – Lefebvre's social and spatial appropriation of Marxist theory, and Turner's apparent anarchist principles of development practice – yet they are known to have no contingent spatial, theoretical, or historical relationship.[1]

The intention of pursuing this apparently unreasonable comparison is to generate a framework of interdisciplinary analysis as a critical lens through which to reveal, interrogate, and contest key moments of intersection between the discourses of Lefebvre and Turner. In doing so this analysis validates the underlying premise that development practices may reflect many of the positive socio-spatial characteristics advocated and aspired towards in Western spatial discourse; albeit in subtle and unexpected ways.

This trajectory of analysis is not simple or straightforward. It first requires some leg work. Our analysis begins with a grounding of Lefebvre's spatial contextualisation of Karl Marx and Friedrich Engels' methodology of dialectical materialism (Fischer, 1973, p. 87). Subsequently, Lefebvre's conception of 'space as a social product' (1991, p. 26) is observed as inherently founded upon the framework of dialectical materialism and the relational processes that produce space (Lefebvre, 2009, pp. 303–305). The principle of space as socially, relationally, and materially produced provides the underlying theoretical foundation upon which our wider research trajectory is built, namely: a critical comparative analysis of the theoretical and practical articulations of dialectic and relational social space as a process and practice.

Turner's discourse remains an explicitly practical and spatial investigation of the social and economic benefits of participatory methodologies and user choice in urban and informal housing (Turner, 1976, p. 153). The apparent anarchist or libertarian principles that appear to drive his practices are never made explicit in his discourse, and only appear in observation, praise, and critiques of his work from secondary sources. Eschewing these prevailing pigeonholings of his work, our analysis will observe how his observations and engagement with alternative housing models in the informal settlements of Peru and the wider Global South can be interpreted as a unique practical contestation of a dialectical and material approach to development. Turner neatly summarised this agenda and agency in his ground-breaking articulations of 'housing as a verb' (1972a). The explicit practicality of his discourse affords us the opportunity to contest a comparison to Lefebvre, and in so doing, to reveal the inherent material and dialectic foundation of Turner's critique of formal housing and his subsequent counter-propositions for participatory socio-spatial development.

In the context of Turner's discourse on urban mass housing and informal set-tlements this chapter also looks to intersect theoretical contestations of 'the city' as a site of critical interdisciplinary comparison in critical Western spatial theory (Harvey, 2003). In the context of Lefebvre's *The Survival of Capitalism* (1976) (and in contrast to predominant structural and political conflations of alterity and illegality (Fernández-Maldonado, 2007, p. 5)), informal settlements and economies of absence can be interpreted as a global urban condition. Thus, upon returning to our comparison of theoretical and practical conceptions of space, Lefebvre's articulation of the inherent contradictions of capitalism and subsequent contesting of 'the reproduction of the social relations of production' (1976, p. 17) provides a further intersection with informal urban settlements as articulations of alternative differential spaces and values (1991, p. 52, 1976, p. 115).

This theoretical articulation of positive yet alternative spatial relations can thus be drawn into critical comparison with Turner's advocacy for housing and development as a progressive and intergenerational process and social practice (1986, pp. 10–12). In contrast to prevailing presumptions of inevitable models of growth, capitalism, and their accompanying political ideologies (Lefebvre, 1996, p. 190), the alternative values, practices, and social relations of informal settlements exist as practicable and socially sustainable examples of the positive implications of heterogeneity and autonomy as a socio-spatial condition (Ingham, 1993; Turner, 1976, pp. 21–23). Thus, if we as architects and spatial agents aspire towards a similar critical advocacy for space as that of Lefebvre, then what can we learn from Turner's (apparently anarchistic) practices and their approach to people and place? And how can these observations, drawn from such different contexts – historical, geographical, social, political, and economical – help us to reframe and re-imagine space and architecture?

Turner's advocacy for housing models based upon networks, autonomy, and heteronomy provides further points of intersection to Lefebvre through a comparison with his advocacy for a spatial politics of autogestion and self-management. Whilst Lefebvre's autogestion offers a positive spatial con-textualisation of the Marxist concept of self-management (2009, p. 14, 1976, p. 40), it equally raises and recognises the dangerous ability of late capitalism to consume and re-appropriate such objects and identities of transgression through co-option and reification (Harvey, 2010, p. 233). When placed in such critical comparison with Lefebvre's theoretical advocacy for autogestion, Turner's practical examples of networked, heteronomous, and alternative development practice are interpreted not as mere aberrations and anomalies of backwards societies (Merrifield, 2006, p. 122). Instead they are here contested as inherently positive realisations of socially produced space, and as a socially and economically logical, contingent, and valid model of grass-roots self-management: a practical and concrete spatial realisation of autogestion?

Subsequently a foundational point of origin for this research trajectory is observed in Turner's contestation of the implications of the central issue of 'Who decides?' (1976, p. 11). Within this simple, eloquent, and critical examination of political authority and hierarchy Turner offers a first connection between the broader spatial, political, and cultural implications of this interdisciplinary comparison. The contestations of hegemony, identity, and values in our later chapters are here contested with Turner's simple observations of the authority and control in the decision making of informal space and their alternative social relations. This simple observation offers a critical lens and framework with which to question and contest the social and political implications of local and global development (Shields, 1999, p. 183), and perhaps to begin to question the limitations of a polarity of state capitalist and socialist ideologies.

In the context of these comparisons, Turner's work can be read anew as a post-structural (and 'Lefebvrian') reinterpretation of development practice and a provocative contestation of difference versus authority, hierarchy versus grass-roots democracy, hegemony versus participation.

Similarly, the intersection of Turner's practices and the principles of dialectical materialism provides a renewed practical agency to Lefebvre's theoretical discourse. In the context of this comparison, participatory and progressive development in informal space can begin to be recognised as a concrete realisation of Lefebvre's articulation of spatial practices: of the notion of social, political, and spatial change as being driven by a dialectical process and explicitly informed and implicated by the concrete material reality of its socio-political context (Goonewardena et al., 2008, p. 100).

A brief introduction to dialectical logic

In order to contest the premise of an interdisciplinary comparison between Turner and Lefebvre it is first necessary to provide a foundational contextualisation of dialectic reasoning. Lefebvre's critique of society and space is based upon a theoretical lineage back to Marx and to Georg Wilhelm Friedrich Hegel.[2] This trajectory of thought itself relies upon the translation of abstract philosophy into a form and methodology that engages with material and social reality. Marx's material and economic re-contextualisation of Hegel's abstract logic is intrinsic to Lefebvre's discourse and is of vital importance to our research premise to directly connect and learn from the comparison of theoretical and practical engagements with space (Lefebvre, 1996, p. 190; Shields, 1999, p. 155).

For Lefebvre the dialectic needed to be grounded in the material reality of space, and thus our premise is to ground this theoretical methodology of dialectical materialism against Turner's realisations of participatory housing development in 1960s Peru.

In essence Hegel argues that ideas[3] are in constant conflict with each other and the result of these conflicts are new ideas. This process in turn leads to new conceptions and new conflicts and so on. This is Hegel's dialectic logic. Much like the classical articulation of dialectics,[4] Hegel's approach contests that whilst 'everything' is composed of contradictions and opposing forces, they are also all part of a continual process of change and evolution (Fichte, 2000; Henrich et al., 2005). Hegel builds upon Fichte, who translated the negativity of Kant's logic of contradictions into a dialectic model. His appropriation of Fichte generates the process structure of thesis–antithesis–synthesis, and in so doing, Hegel is widely acknowledged as providing the bridge between Kant and Hegel (Shields, 1999, p. 11; Stepelevich, 2008).

Crucially, this thesis–antithesis–synthesis process is not repetitive, but iterative. Change was therefore a continuous dynamical process and helical not circular (Hegel, 2010, p. 46). The implication of Hegel's logic is its inherently positive identification of contradiction, mediation, and negotiation as a process that led towards synthesis.[5] Politically contextualised by Marx and later and spatially contextualised by Lefebvre, the premise of a positive dialectical approach to space is intrinsic to our reseach.

For clarity we must define a theoretical baseline for the study of the practical and material implications of the concept of dialectical materialism. In this case we shall begin with Engels' discourse on industrial Manchester published in 1844 (Engels, 1844; Engels and Marx, 1975), and Marx's first political and economic works that emerge at this key point in history.[6] Here a critical intersection emerges between Hegel's dialectical logic and the discourse of Marx, who appropriates and re-tools dialectics for use as an analytical method to contest the socio-political and economic conditions of the nineteenth century (2013, pp. 15–16). Yet at first Marx was dismissive of Hegel's abstract and inherently negative articulation of dialectic logic, specifically contesting the philosophical abstraction and internalised contradiction of the dialectical model as defined abstract–negative–concrete.[7] Lefebvre's treatise on the dialectic similarly contests the same sense of injustice at these structural abstractions and their persistence a century later:

> Hegel was not content merely to deepen the content and make it explicit in order to attain the form, he reduced it to thought, by claiming to grasp it 'totally' and exhaust it. He insists on the rigorously and definitively determinate form which the content acquires in Hegelianism. All the determinations must be linked together in order to become intelligible. As far as Hegel is concerned, these connections are not discovered gradually, obtained by an experimental method; they are fixed.
>
> (cited in Shields, 1999, p. 51)

In contrast to the absoluteness and fixity of abstract thought, Marx and Engels would together provide a paradigmatic contribution to the dialectics of philosophy, sociology, and economics, through their observations and critiques of the implications of the industrial revolution for the common man (Engels and Marx, 1975, pp. 295–296). Their accompanying critique of Hegel reflects a collective outrage at what they perceived to be the abstract isolation in which Hegelian philosophy existed. Hegel's derivation of a form of pure abstract philosophy was for them an 'esoteric history of the abstract mind – alien to living men – whose elect is the philosopher and whose organ is philosophy' (cited in Lefebvre, 1968a, p. 79).

This critique of Hegel's dialectic method came to define and give critical validity and purpose to Marx and Engels' struggle to grasp and engage in the relational and material context of space (Fischer, 1973, p. 152). It is insightful here to note how the comparisons explored in our research resonate with these innovative critiques of political and economic realities as interdependent with material and social contexts.[8] Thus years after rejecting Hegelianism, Marx describes salvaging the process of dialectic reasoning as a kernel of logic that he described as 'the only valid element in the whole of existing logic', by standing Hegel on his head (cited in Lefebvre, 1968a, p. 84).

> The dialectic method, worked out first of all in an idealist form, as being the activity of the mind becoming conscious of the content and of the historical Becoming, and now worked out again, starting from economic determinations, loses its abstract, idealist form, but it does not pass away. On the contrary, it becomes more coherent by being united with a more elaborate materialism.
>
> (Lefebvre, 1968a, p. 84)

In reaction to the social inequality observed in industrial Manchester, the UK, and industrial Europe, Marx and Engels appropriated Hegel's dialectic process and contextualised it within a concrete and materialist field of discourse (Fischer, 1973, p. 81). In contrast to the abstraction and internal negativity of Hegel's logic, this critical analysis would place the relationships between things, people, and place at the crux of a logical engagement with the social, economic, and political contestation of the inequalities of industrialisation (Lefebvre, 1968a, p. 98). Marx's historical materialism utilises the dynamic of idealism (of Hegel's interpretation of history as trajectory towards reason and hence freedom) and the conditioning stated by materialism (as an interpretation of Ludwig Feuerbach)[9] and fuses them to reveal something new: the proposition that we are conditioned by our environment, but we can intervene to change these conditions that affect us precisely because time unfolds in a socio-material and historical evolution (Lefebvre, 1968a, pp. 120–121).

Change is possible (whether it be *positive* or *negative*) simply because the world is not abstract, but materially conditioned. Whilst this in itself might not seem controversial, Marx realised that if every idea, practice, and social relation is constantly changing then no condition is natural, inevitable, or fixed – they are made. In the context of Marx's observations of social inequality and the political ideology of the mid nineteenth century, dialectic logic was re-purposed to contest not merely Hegel's abstract philosophy, but the material and economic reality of industrialisation itself.

The intersection of historical materialism and dialectic logic became Marx's method of exposition. It formed a new way of seeing, valuing, and contesting the material reality of spatial content (Fischer, 1973, p. 157). Consequently the first thing to look at in understanding how a society works is to look at the things – products, housing, social relations – it produces, and how (and ultimately why) they are produced (Fischer, 1973, p. 53).

Whilst the discussion above is by necessity a somewhat expedient exploration of the origins of dialectical materialism, its significance to Lefebvre's discourse (Elden, 2004, p. 33) and the premise of our wider research cannot be over-estimated. It has been important to explicate the material and social foundations of Marx's logic before proceeding with the critical comparisons of purposefully practical (Turner and Hamdi) and theoretical (Lefebvre and Massey) prota-gonists. Marx's discourse provides explicit validation to the re-appropriation of the principles of dialectical materialism in order to enable a critical comparison to the material and practical contexts of Turner's participatory development. It also highlights the social imperatives and contestation of inequality that drove the work of Marx, Engels, and Lefebvre as a trajectory that continues into the later works considered by this book of Massey and subsequently Chantal Mouffe, Ernesto Laclau, Homi K. Bhabha, and Gayatri Spivak.

Lefebvre's dialectical materialism

Lefebvre's discourse *Dialectical Materialism* (1968a) narrowly preceded his more famous work *The Production of Space* (1991) and is a distinctly different text, offering a short, focused analysis of Marxist logic that would come to be extensively utilised in much of his later pioneering works (2009, pp. 32, 40, 2009, pp. 303–305, 1991, pp. 129, 417). In exploring dialectical logic Lefebvre found the embryonic framework of an explicitly spatial methodology by beginning to interpret space as relationally constructed in a continuously evolving process (1968a, p. 120). This spatial turn informed Lefebvre's use of dialectical and material reasoning as a critical lens and observational method.

The insights he drew from this relational analysis of space as a product prompted Lefebvre to transcend the institutional Marxist interpretations that he considered as pervading much of his academic contemporaries (2009, pp. 100–106, 1980, pp. 19–32; Merrifield, 2006, p. 4). In direct criticism of a prevailing institutionalised ideological Marxism, Lefebvre proposed that Marx had to be understood as a spatial 'programme or project [which] must be brought face to face with reality, that is with the praxis (social practice), a confrontation which introduces new elements and poses problems other than those of philosophy' (Lefebvre, 1968a, p. 19). This agenda of engagement with the material and social reality of spatial conditions and relations is crucial to our study. It is a theme that recurs throughout our research trajectory in both theoretical and practical points of comparison.

If Marx can be said to have rescued the dialectic method from Hegel's abstract philosophy, then Lefebvre is equally valuable for his attempt to salvage from dialectical materialism the political imperative found in the notion of spatial practice and praxis (Merrifield, 2006; Shields, 1999, p. 152). Utilising dialectical materialism Lefebvre describes praxis using the language of movement, conflict, and contradiction. Within this utilisation of movement is a tacit implication of spatial practices with the idea of a continuum of space and time provoking change through the praxis (Lefebvre, 1968a, p. 94).

Thus Lefebvre's articulation of spatial practice and social relations as interdependently linked by praxis implies an intersection with dialectical space, process, and evolution, and with our comparisons in this chapter of Turner's discourse of participatory development practice. As we will see, his models of progressive housing based upon informal settlement practice methodologies explicitly implicate the production of space and social relations with grass-roots, heteronomous, and networked social relations that empower social, economic, and institutional change. Thus, Turner's spatial and concrete observations of such participation can be seen in a valuable new perspective if critically re-read in comparison with Lefebvre's relational space of dialectical materialism:

> Practical activity and effective action is what we and existence are all about. As well as being stimulated by them, actions lead to problems. And problems raise issues. Issues, in turn, indicate principles for action, while principles determine the resolution of issues. And finally, principles are guides for practice as well as being generated by it. These elements in the development of a process for action must be fully recognised for any coherent discussion of social, institutional and environmental change.
>
> (Turner, 1976, p. 103)

This comparison of Turner's work as a dialectical materialism offers a re-reading of his practices in Peru as advocating the same aspirations for space that Lefebvre was exploring contingently on the other side of the world in Paris. Lefebvre's and Turner's respective critical discourses lead to a conception of space as inherently both materially and relationally constructed, whilst also implicating further clear interdependent connections to concepts of identity and culture:

> The praxis is where dialectic materialism both starts and finishes. The word itself denotes, in philosophical terms, what common sense refers to as 'real life', that life which is at once more prosaic and more dramatic than that of the speculative intellect. Dialectical materialism's aim is nothing less than the rational expression of the Praxis, of the actual content of

life – and correlatively, the transformation of the present Praxis into a social practice that is conscious, coherent and free.

(Lefebvre, 1968a, p. 112)

It is the idea of a conscious, coherent, and free social practice we will now move on to discuss, and we will explore the examples and concrete realisations of dialectic materialism found in John Turner's participatory model of housing praxis.

User choice participatory housing

Between 1957 and 1965 Turner lived and worked predominately in the rapidly expanding urban squatter settlements of Peru for independent and government housing agencies in the promotion and design of pro-poor community action and self-help housing.[10] In comparison with Lefebvre's spatial critique Turner's practices, discourse, and observations of housing offer points of intersection and resonance. In particular, these critical comparisons highlight his observations of the necessity of user choice and participation in mass housing as a form of spatialised dialectic materialism, and as specifically interpreting, questioning, and engaging with concrete social and economic content. His critique of the abstract and elitist nature of architectural practice can be observed and compared as aligning with Marx's critiques of Hegel's abstract dialectic logic, and subsequent engagement with the real life implications of material and economic contexts. Thus, Turner's articulation of the underlying conflict between his practical confrontation with space and his education and profession as an architect are implicit within the contextualisation of his retrospective discourse:

It was only after living and working in Peru that I began to articulate the dissatisfaction shared with so many contemporaries. We felt and knew that architecture cannot be practiced as if it were an independent variable – as though the architect had no social or political responsibilities – yet neither could we accept the marxist antithesis. It seemed as absurd to believe that social structure could be changed through architecture as it was to believe that architecture should be entirely subjected to the official interpretation of taste.

(Turner, 1972b, p. 123)

Trapped between abstract architectural formalism and the institutionalised state Marxism, Turner's words resonate with an underlying premise explored throughout the comparisons explored in our research. As outlined in this quote Turner's practical and hands-on engagement with a developing world context led him to a critical interpretation of the socio-political engagement of his architectural contemporaries (Baird, 2003, pp. 265–272). In contrast with the declining ideologies of CIAM (Congrès International d'Architecture

Moderne) (Hughes and Sadler, 1999, p. 29), Turner utilised a broad context of political and sociological theory (Ward, 1972, pp. 8–10). Perhaps most notable are his readings of the anarchist politics of Peter Kropotkin and Ivan Illich, Giancarlo de Carlo's problematisation of housing (1949) and Patrick Geddes' general systems theories (Geddes, 1949). He sought an active engagement with a broader interpretation of architectural context as being interdependent with political, economic, and human relationships (Turner, 1986, pp. 24–25, 1967, p. 179).

This desire to engage in a broader and relational context of architecture and development provided the beginnings of Turner's exploration of what this research contends is a dialectical approach to the materialist reality of space. In this context Turner's appropriation of Geddes advocacy to 'involve himself as closely as he could with all the people concerned' (Fichter and Turner, 1972, p. 122) resonates with the same materialist social analysis and advocacy of Marx and Engels. Yet crucially Turner's discourse is not limited to political observations, social discourse, or economic theory, but is contested in spatial practices of development and the concrete reality of informal settlements and mass housing. It is this explicitly spatial turn of Turner's work that defines the comparison to Lefebvre's spatial re-appropriation of dialectical materialism and critical observations on the urbanisation of France (Lefebvre, 2003, pp. 126–130).

This distinction between the practical grass-roots experiences of Turner, and the theoretical engagement with politics and state governance by Lefebvre is crucial. It defines their approaches – one explicitly practical, and one theoretical – as seeking to engage with space and spatial relations in two distinct directions: Turner building from the ground upwards (an anarchistic approach) and Lefebvre seemingly seeking a form of top-down state socialism.

The rapid urbanisation of Peru provided a context for Turner to confront and contest the problems and potential of mass housing and social inequality. His major contribution to this field marks a contestation of the contradictions of the top-down models of housing that he observed in South and Latin America (Dovey and King, 2012, p. 291; Harris, 2003, pp. 247–251). This analysis is exemplified by the social and economic disjunctions between the negative social effects of state-sanctioned *superbloques* housing and the positive social potential of informal settlements in the urban peripheries that were generally assumed to be illegal, socially detrimental, and valueless ('A Basic Issue: Values and Standards', in Turner, 1974).

Turner observed across Latin America and the wider developing and urbanising world that the principles of modernist housing were being advocated and rapidly imposed upon cities by government-sanctioned centralised and administered housing programmes ('Housing by Trained Professionals for Untrained Masses', in Turner, 1974). In contrast to informal settlements Turner critiqued these housing programmes as generating economic relations and social spaces of alienating. Crucially he also recognised that this was not simply because of their abstract form and planning but also because they separated people from the participation and production of their housing and values.

Treating housing and people as quantifiable and economic values created diseconomies and dysfunctions of social products, uses, and values.[11] The scale and homogeneity of formal centralised housing development provides quantitative and bureaucratic solutions that are intrinsically unable to adapt to fit the variety of lifestyles that are vital in the economic evolution and social sustainability of cities. Furthermore, Turner is incisively critical of the disjunction of central and abstract models of housing as socially alienating and divisive, recognising that this is further compounded by the relationships such practices produce between all concerned and the environment.[12]

The formal standardisation of modernist and symbolically Westernised housing models (and space) was implicitly dependent on economic models of production that benefit a scale and homogeneity that have two main effects (Turner, 1976, pp. 46–47). Firstly, they isolate the economic benefits of manufacture in the hands of large corporations, restricting the opportunities for relations of small and medium businesses to compete. Secondly, because of their alienation and abstraction from the actual users of housing they inherently generate spatial misfits of design and economy that are targeted precisely at the part of the population who can least afford such inappropriate waste (Turner, 1976, p. 51). In contrast to this, Turner advocates an alternative understanding of housing as defined not by economic and political quantification of what it is, but by quantitative and heteronomous contestations of values in what housing does:

> If the usefulness of housing for its principal users, the occupiers, is independently variable from the material standards of the goods and services provided as the case studies and other sources show, then conventional measures of housing value can be grossly misleading. As long as it is erroneously assumed that a house of materially higher standards is necessarily a better house, then housing problems will be mis-stated.
>
> (Turner, 1976, p. 60)

In the 1950s and 1960s the widely accepted response to the informal settlements on the edges of cities. spatially, socially, and economically, was to utilise state intervention to impose the stability and rigidity of a formalised model (Turner, 1972b, pp. 143–144). For Turner this presumption of the social and economic benefits of formal, centralised, and modernist housing interventions is based upon a naïve and prejudicial misconception that people in informal settlements are unable to make rational judgements about their own space and everyday lives for themselves (1972b, p. 141). The evidence of this is supposedly demonstrated in the informality of their habitation and interaction as individuals and a community beyond normal conventions. These political assumptions act to validate a direct imposition of control and authority by formal, centralised state housing and a rejection of any positive potential of self-reliance, self-governance, and social sustainability that might exist within the dialectical materialism of informal development. In contrast Turner's celebrated observations and interpretations of this context were some of the first attempts to demonstrate that the exact opposite is true (1972b, p. 141).[13]

This observation coincided with the ground-breaking work of anthropologist William Mangin, who would become a key academic contemporary of Turner in Peru. In 1967 Mangin published a paper in the *Latin American Research Review* titled 'Latin American Squatter Settlements: A Problem and a Solution' (1967), within which he exposed the unwarranted social stereotypes of irregular settlements. He concluded that given moderate and sustained support through self-help, mutual aid, and localised support such settlements offered demonstrably better social value as models of intergenerational development over periods of fifteen to twenty-five years (1967, pp. 74–75).

In this theoretical context Turner's advocacy for housing consolidation and the self-help progressive development of informal settlements aligns with the observations of Charles Abrams (1964, 1966). Abrams and Turner similarly advocate that given the economic incapacity and social homogeneity of the government and the formal housing market self-help was an acutely appropriate and intellectual response by the urban poor themselves to provide housing at an affordable price and on a large scale. Here Turner explicitly observes the critical paradox that 'governments have done so little with so much, whilst poor people have done so much with so little' (1979, pp. 1135–1146).

In contrast to government backed 'modernist' projects the progressive development of urban migrants who appropriated land either by illegal squatting or informal purchase could be seen to generate sustainable social and economic improvement over time usually in the form of grass-roots community action. The organisation and collaboration of people to level and maintain streets, hook up rudimentary services and electricity distribution, and eventually to agitate for local state services was both economically valuable, but more importantly was socially conducive to sustainable communities (Ward, 2008, p. 290). In the context of political incapacity and economic absence Turner revealed informal settlements and progressive development as simply an 'architecture that worked' (1968).

Having introduced positive existing connections to Mangin and Abrams (and our own proposition of a comparison with Lefebvre) it is necessary here to note various critiques of Turner that exist within academic discourse. One critique pertains to whether Turner acknowledged clearly enough the existence of self-help housing prior to his interventions in Peru. Harris is explicitly critical of this supposed 'deafening silence' in spite of the time gap between the first examples of self-help in the 1940s and 1950s and its resurgence in the late 1960s (Harris, 1998, 1999). This critique relates to Jacob Crane's work on self-help at the Housing and Home Finance Agency (HHFA) in Washington in the 1940s, where he coined the phrase 'self-help' in 1945. Harris also highlights the paradox that although the principles of self-help were developed in the USA they were not actioned in their country of origin. Instead, they were first enacted in Puerto Rico in 1939 in a project fully funded by US federal government and international development agencies.

This projection of policy from Global North to Global South, and the accompanying financial support for people living elsewhere is a practical subject and socio-spatial concept of that we see recurring repeatedly throughout our wider research premise. The principles of self-help housing and the policies of the US-backed HHFA spread out across the rapidly developing and urbanising countries of Latin America and rose to prominence later in the work of key housing specialists like David Vega-Christie in Peru (1948). His work in turn led Eduardo Neira (an architect at the Peruvian Ministry of Public Works) to establish a pilot project with squatters in Arequipa, and to invite John Turner to act as an advisor on the project. Writing retrospectively Peter Ward notes Turner, Mangin, and Abrams as having not recognised that rather than inventing self-help they merely introduced it to a wider audience, for Ward a case of 'putting old wine in new bottles' (2008, p. 290).

Yet in spite of these points it is indisputable that from within the informality of Lima's *barriados* Turner observed and pioneered a new (and simple) way to appreciate and value the methodological approaches and principles that define informal settlement:

- Firstly, that irrespective of the material appearances of the results, people are almost always the best judge of their own needs and actions.
- Secondly, that by taking charge of their destinies, people and communities are able to generate models of appropriate, reactive, and sustained development as a logical response to a context that cannot be understood in abstraction – a process that this chapter contends can be described as a materialist dialectic.
- And thirdly, that through the continuous process of progressive development, the social and economic circumstances of informal settlements should start to be viewed as the answer to economic deprivation instead of the problem itself. (Turner, 1976, pp. 137, 140, 149)

In this context Turner's socio-economical observations and practical realisations of alternative development explicitly advocate the social and political importance of autonomy, choice, and the freedom to build:

> When dwellers control the major decisions and are free to make their own contributions to the design, construction or management of their housing, both the process and the environment produced stimulate individual and social well-being. When people have no control over, nor responsibility for key decisions in the housing process, on the other hand, dwelling environments may instead become a barrier to personal fulfilment and a burden on the economy.
>
> (Fichter and Turner, 1972, p. 241)

Turner's critique of the cultural and economic implications of governmental control and architectural authority crucially coalesce here with broader political implications of participatory and user-informed housing to potentially represent something more than the sum of their individual parts.[14] When engaging in rich and vibrant cities of both formal and informal settlements it *must* be inherently more valuable to empower, facilitate, and advocate for people and communities to produce places for themselves in a model of intergenerational and progressive growth (Turner, 1976, p. 17).

This underlying principle of people having freedom, opportunity, and control so that they might build for themselves is both statistically, economically, and practically validated by Turner's observations (1976, pp. 66–70), but is also emblematic of a deeper recognition of the need to pursue alternative social and political contestations of value (1967, p. 179).

> It seems that all national and international housing and planning agencies, mis-state housing problems by applying quantitative measures to non or only partly quantifiable realities. Only in an impossible world of limitless resources and perfect justice – where people could have their cake and eat it too – could there be a coincidence of material and human values. [...] So long as this fact of life remains, and as long as people's priorities vary, the usefulness of things will vary independently of their material standard or monetary value.
>
> (Turner, 1976, p. 61)

Such a political advocacy for the value of user-defined housing can be likened to Lefebvre's observations that organisations tend to institutionalise the space and values of everyday life, which leads to social alienation and the reification of such activities (1969, pp. 67–68, 2009, p. 116; Merrifield, 2006, p. 47). These observations also suggest an explicit comparison with Lefebvre's contestations of use and exchange value in the articulation of social relations and production of space (2009, pp. 191–194). As with Turner's observations of the social alienation of formal housing as an institutional product, Lefebvre critiques the world of commodities and exchange value as generating its own reductive logic, with use value resigned to mere sign and symbolic exchange, noting that this 'is a world which de-dialectises itself, defusing contradictions and conflicts' (1976, p. 71).

In his confrontation of these issues Turner articulates practices that support and advocate informal and user-choice housing models; development practices that engage and contest the contradictions and conflicts of space dialectically through autonomous and progressive growth. For Turner, the socio-spatial and political practices of informal housing contest the social and economic *value* of housing as being interdependent with the choice and participation of its users (Fichter and Turner, 1972, p. 159).

This spatial and relational turn represents a form of material dialectic reasoning, which recognises the necessity of working in close proximity with the social and material reality of space. In his contestation of the potential value of informal housing Turner explicitly acknowledges material and relational contexts (1972b, pp. 124–125). He does so by facilitating participatory and socially innovative practice that transcends architectural preconceptions. His analysis not only explores the issues that frame the delivery of much of our urban environment, but goes further. In his critique of the socio-economic context of informal settlements he was able to propose and realise concrete alternatives that demonstrate empirically that it is a more socially responsive and economically viable practice. Thus, it is revealing to re-consider Turner's practices against Lefebvre's socio-political critiques of space, and reciprocally, to re-consider Lefebvre's discourses in the context of Turner's apparently anarchistic spatial practise. In order to confront and contextualise this argument further we will next introduce an example drawn from Turner's analysis that demonstrates this contestation of value, choice, and necessity.

Supportive shacks and oppressive houses

The theoretical premise of our research is built upon re-reading and re-valuing simple examples such as this one from Turner. Whilst they are at times difficult to connect with the reality of a Western context, looking past the contradictions offers valuable opportunities for critical reflection on the values and ideas that define space and architecture.

Turner's analysis of and advocacy for both the social and material efficiency and the relational sustainability of informal housing settlements is best surmised in his analysis of what he describes as 'supportive shacks' and 'oppressive houses' (1976, pp. 52–53). This comparison forms part of a detailed social study of a range of twenty-five examples from urban Mexico which each describe a spectrum of material and social values in their individual situations (1976, p. 59).

In contrast to the presumptions of prevailing large-scale housing developments Turner's analysis of these examples focuses upon the relative social values of both formal and informal housing. This critical analysis and documentation of the three spatial criteria – *tenure, security*, and *access* – provides quantifiable evidence that the rich heteronomy offered by the community networks of informal housing achieves a social efficiency that cannot be achieved by homogenous centrally administered housing (1976, pp. 68–69). His study utilised a cross-sectional analysis to look at multiple examples that accurately reflected the alternative social criteria, material reality, and values of dwelling in informal settlements. Subsequently, Turner explored the social significance of simple things like tenure, security, and access in the supportive shack:

> All these conditions are met by the car painter's shack. While the family would undoubtedly enjoy a higher standard dwelling this is relatively unimportant. [...] This materially very poor dwelling was extremely well located for the family at that time; the form of tenancy was ideal, giving

them security without commitment and the freedom to move at short notice; and the shelter itself provided all the essentials at minimum cost. The shack was, therefore, an admirable support for their actual situation and a vehicle for the realisation of their expectations.

(Turner, 1976, p. 58)

Within these observations and the wider study is an explicitly material and dialectic methodology of logical analysis. Rather than relying upon assumptions or ideologies of housing and growth Turner studies the choices made by people facing the reality of necessity in order to understand and interpret their specific value (Fichter and Turner, 1972, pp. 164–168). The vast potential for mismatches offered by formal settlements became clear in the contrasting example of the oppressive house:

The mason's modern standard house is disastrously unsatisfactory. [...] This family now lives in a vastly improved modern house, equipped with basic modern services and conveniences. However, this 'improvement' is endangering the lives of the family members, and in human and economic terms has led to a dangerous deterioration of their condition. Incredibly, the family is required to pay 55 per cent of its total income to meet the rent-purchase and utility payments.

(Turner, 1976, p. 56)

It should be noted that Turner takes pains to not simply dismiss the value of the more materially substantial housing that the state sought to offer. This is not an implicitly anti-capitalist or anti-state analysis of housing. Turner's work explicitly recognises the potential for the state to help and facilitate the improvement of informal settlements in his advocacies for locally administered 'sites and services' programmes (1986, pp. 6–8; see also Dovey, 2013; Schon, 1987). Yet these limitations and problems of prescriptive formal housing cannot be ignored:

In their previous situation there was a positive match between their priorities and their housing services. The family's housing priorities were naturally for security of tenure and access to their sources of livelihood. [...] They were therefore able to maintain their rudimentary but tolerable shack in order. They were able to feed and clothe themselves reasonably well, and most importantly, they could save for security in their old age. In their present situation they have lost nearly all of these advantages and they acquired others of secondary importance. They lost access to a major source of income and as events proved, were unable to maintain the absurdly high level of housing expenditure. [...] Whether this family was more comfortable or not, with the anxiety and hunger that they certainly experienced as soon as their savings were used up, is a not-so-open-question.

(Turner, 1976, p. 59)

Formal and informal housing exist on a spectrum of services and choice that adapts and evolves over time to the needs of the people. However, in contrast to housing as a product of intervention, Turner's alternative advocacy for progressive self-help housing development programmes is specifically designed to counter social, political, and economical mismatches. By valuing and advocating the notion that people themselves are best placed to judge the best solution to their own situation, Turner's observations critique the paradox of the false social values inherent in formal housing both in the context of economies of absence and beyond:

> Some of the poorest dwellings, materially speaking, were clearly the best, socially speaking, and some, but not all of the highest standard dwellings, were the most socially oppressive.
>
> (Turner, 1976, p. 52)

In light of Turner's critique of formal and informal housing, the comparison to the dialectical materialism of Henri Lefebvre can now begin to be articulated more clearly. By re-contextualising and re-reading these examples in relation to each other, it becomes clear that Turner's work is explicitly a practical critique of the material, economic, and social relations that defined the housing in 1960s Peru. The practices, process, and space of Turner's housing advocacy for the value of informal settlements and housing can thus be considered as realisations of Lefebvre's articulation of space as a process of dialectical materialism. Turner's observations of the false assumptions of high and low quality housing also begin to offer the first points of reflection upon the opportunity to learn from such practices in comparison to the architectural practices of Westernised space and the Global North.

Other recent contemporary developments also help to highlight the intersection of informal housing in the South and a re-imagining of space and the methodologies of architecture. The work of UTT (Urban Think Tank) and its intersecting SLUMLab (Sustainable Living Urban Models Laboratory) project have emerged in recent years as perhaps the most notable agents and advocates of engagement in alternative city spaces and informal housing. UTT's work on Torre David in Caracas Venezuela has provided a highly popularised and much discussed example of the confrontation of architecture and alterity (Brillembourg and Klumpner, 2012). The informal/illegal occupation of a building intended for high-rise business and commercial occupation has become one of the most recogniseable representations of the confrontation of alterity and architecture.

A more explicit reference to Turner's work can be seen in SLUMLab's recent design and production of a project entitled 'The Empower Shack', a prototype for South Africa's informal slum areas, which is essentially a template for urban densification and upscaling. The connection to Turner's description and advocacy for the positive potential of 'supportive shacks' is palpable. UTT's simple design is an example of global cooperation[15] and engagement with vital questions that face the rapidly urbanising Global South.

Much of the critical thinking, intentions, and aspirations for this project reflect Turner's original thinking. The relationality of the architectural agency which underpins the empower shack responds at a local scale to the economic absence of informal communities in a positive and open way, seeking to facilitate and support plausible housing goals that can be repeated and built upon in the future (Jiron, 2010). Yet, for better or worse it is also a globally connected project, and reflects a relational contingency of international questioning of the post-colonial fallout still felt throughout the Global South.

In spite of the need to remain critical and objective about these projects in the complicated context of the politics and economics of global yet post-colonial space,[16] examples like these begin to reflect a re-imagination of architecture as a relational response and representation of social intelligence and honesty. It suggests a re-invigoration of architecture with the agency and purpose needed to respond to a relational understanding of space, and as a positive confrontation with the reality of multiplicity and other people's lives and spaces. It questions the politics, economics, and identity of architecture by engaging with alterity in an open dialogue that attempts to learn collectively through collaboration with other people's lives and values.

The social, political, and economic implications and relations of these examples are all too easily lost in their reduction to sensationalism of visual imagery and festishisation of informality as poverty (Dovey and King, 2013). Yet when considered with the critical lens of relationality the deeper social agency and political implications of such projects become far more valuable. As such the comparisons in this book seek to frame examples principally from Turner but also the likes of UTT within a wider theoretical discourse and discussion of the implications of alternative and informal space in the Global South.

The critical concept and implication of dialectical materialism is of vital importance in any attempt to contextualise the implications and positive potential of an alternative re-imagining of space. By learning to interpret spaces of informality in the Global South the critical lens of dialectical materialism can subsequently begin to be turned inwards to critique the political, economic, and social foundations of Westernised architecture and space. From this perspective, informal space and housing is a cypher with which to critique and question our assumptions of the relative social value of architecture and space.

The contrast of the architectural methodologies that produce projects like 'The Empower Shack' to conventional Westernised architecture and space is on face value too large to maintain valid critical purpose in a Global North context. How can such projects possibly offer a critique of the architectural requirements of Westernised space and lives? Yet in examining these projects with the critical lens of dialectical materialism and a Lefebvrian interrogation of space, the implications become quite stark. Despite being separated by five decades of time and thousands of miles of geography, projects like these inherently exist and flourish because of the contexts of extreme economic absence and austerity. The fact remains that the economic conditions that define such projects are a global construction that continues to persist in the global periphery,[17] and that remains an explicit yet culturally overlooked implication of the globalised economy and the aftermath of colonial space and identity (see variously: Coetzer, 2010; Kusno, 2000; Nalbantoglu and Wong, 1997).

Yet there also remain opportunities to take positive solace from the existence of Turner's and UTT's work. These alternative approaches to housing reflect an engagement with and realisation of different values, spaces, and socio-spatial methodologies to the conventional presumptions of Westernised space. In essence, the shift towards an alternative set of values and purpose of architecture is linked to the social relations that produce space, and offers the potential to re-imagine our own space by looking at the space of others. Housing that is chosen and defined (and maybe also built) by the home-owners themselves is evidently more efficient than centrally delivered housing. The implications of this simple observation are hard to overestimate.

The only challenge remains to learn to critically understand the relation-ships that underpin projects like these and to critically learn from such simple and humble alternative approaches to people and place, money and value, identity and choice. Perhaps the clearest example of architects engaging with this question in the UK and wider Westernised space are Architecture 00 and their open-source WikiHouse project. Applying knowledge learnt from Turner to projects like this and the wider social and political relationships of Western space might then allow us to begin to piece together the potential for our aforementioned idea of a re-imagining of architecture and space.

Housing as a verb

As we have already highlighted, Turner's observations of informal settlements confronted and critiqued the conventional interpretations of the value of housing and ownership of land as purely economic factors.[18] In contrast to the prevailing tide of modernist ideology and abstract utopianism he docu-mented both the economic and social efficiencies in facilitating informal housing as a progressive and logical social process that contrasted positively to formal mass housing interventions (Turner, 1986, pp. 8–9). These simple yet profound observations are Turner's clearest contribution to development theory and practice: a re-valuing of participatory and grass-roots housing methodologies as the most socio-economically efficient and sustainable means to house people across the world (1986, p. 14). Across his various written discourses Turner demonstrates the empirical, practical, and social possibilities of his alternative progressive approach to valuing and supporting informal housing (1976, p. 64).

By contesting the values of centrally administered and hierarchical housing Turner recognised informal settlements as being invaluable opportunities to observe and learn the practical implications and possibilities of non-hierarchical housing (1976, pp. 37, 46). The broader political implications of such observa-tions become apparent because Turner used his analysis to inform alternative development methodologies, practices, and discourses in which he advocated the political and economic cooperation and support of informal and grass-roots housing settlements (1976, pp. 127–140). As Peter Ward observes, in contrast to prevailing political ideologies of instantaneous development Turner's advocacy to support existing informal sites and communities reflects a controversial need to actively engage with alternative people and practices as a potentially positive solution to the urbanisation of cities (2008, p. 305; Turner, 1972a, p. 152). This analysis combined practical and situated analysis of the material context of informal settlements (and more specifically the *barriados* of Lima Peru) with a broader political and economic critique of projected Western values:

> As the cases show, the performance of housing, i.e. what it does for people is not described by housing standards, i.e. what it is, materially speaking. Yet this linguistic inability to separate process from product and social value from market value is evident in both commercial and bureaucratic language.
>
> (Turner, 1976, p. 60)

For Turner it was imperative to also speak of the social and human value of housing as a social process, and it was this belief that lead to his innovative critique of the assumption that housing is a noun – a unit of measure for the stock of dwelling units – instead of perceiving it as a verb – a social process (1972a, pp. 148–149). This alternative interpretation of housing sought to value, support, and advocate the freedom of people to build housing and communities by themselves and is an implicit contestation of hierarchical and ideological Western development methodologies generally imposed on the developing world (Bauman, 2000, pp. 59–60; Esteva, 2010; Sachs, 2010). Turner realised that the practical reality of Latin American urbanisation and informal settlements was a materialist paradox to Western quantifiable values and standards:

> The obvious fact that use values cannot be quantified worries those who assume that housing can only be satisfactorily supplied by large-scale organisations. The immeasurability of use values is not in the least perturbing to the conventional capitalist. His value system can only admit the existence of market values in the sphere of commercial production, distribution and consumption.
>
> (Turner, 1976, p. 65)

In complete contrast to the assumption of top-down, centrally and institutionally administered housing Turner advocated supporting people to house themselves to their own needs and requirements (1972a, p. 169). The socio-economic reality for people living and working in urban squatter settlements suggested an antithesis of housing that isn't derived from the aspiration of a Western ideology, but from the material reality of the context. In this comparison with Lefebvrian spatial agency, Turner's was an interpretation of development and housing not as a noun, object, or product, but as a process, practice, and verb (1972a, p. 175).

Perhaps the most noted of these practical methodologies for progressive housing was Turner's advocacy of 'sites and services' programmes (1983, pp. 2–3). In such programmes a balance was met between the state providing basic housing sites, roads, and services within which urban migrants could readily appropriate and self-manage the space for themselves (Schon, 1987, p. 361). Over time such sites were upgraded through mutual cooperation from both government and individual action. This principle was also widely applied to existing informal settlement upgrading programmes (Peattie and Doebele, 1973, p. 67).

The conception of housing as a verb is an implicit engagement with a process of self-help as a leveraging of social capital. Whilst this idea of social capital was not popularised until the 1990s by Robert Puttnam (Puttnam et al., 1993), Peter Ward suggests that the idea was implicit in Turner's advocacy of the social capabilities of informal settlements. Furthermore, Ward sites the potential origins of self-help housing within the community planning efforts of 1950s London, which appears to generate a similar paradox to our comparison of Lefebvre and Turner: the theoretical planning idea of self-help housing is perhaps gestated in a Western space and context, yet is now abstracted from its origins and only exists in developing countries (Ward, 2008, p. 289).

The co-option of the principles of user-defined housing development can be simplistically interpreted as their transaltion into concepts of 'self-help' and subsequently 'sweat equity'. The neoliberal economic and political policies that drove the emerging global institutional powerhouses of the UN and World Bank at first only slowly acknowledged the work of practitioners like Jacob Crane in 1940s Puerto Rico. Yet twenty years later in Peru the work of David Vega Christie and John Turner would become institutionally appropriated, adopted, and subsequently co-opted by the increasingly neoliberal political and economic agendas of global NGOs (Ward, 2005).

The supposition Ward draws from this paradox is that the promotion of Turner's development methodologies by the UN and World Bank has nega-tively interconnected the idea of grass-roots self-management and self-help with the image and preconception of poverty, isolating it as a planning model only suitable for developing nations (Sanyal et al., 2008, p. 17).

Turner himself would later refute the notion that his work and the principles of user-defined development could be reduced and de-politicised to merely the neoliberal economics of 'sweat equity' (Turner, 1992). Thus our comparison of Lefebvre's dialectic process to the notion of housing as a verb posits a renewed intersection of planning and spatial critique in the disparate contexts of the Global North and South. The social and economic contradictions of state intervention housing are logically negated and mediated by Turner's analysis, before being transcended and re-articulated as a spatial synthesis in his advocacy for the solution to be found in the social capital of informal housing (Turner, 1972a, p. 72). Thus the inherent relational and material foundation of this analysis is eminently comparable to the political articulations and contestations of Lefebvre's dialectical critique of *The Survival of Capitalism*. For Lefebvre space is the medium in which the social relations of reproduction are contested in developed and developing countries alike (Merrifield, 2006, p. 153). Whilst the question of how to apply Turner's practices to a Western context remains, the positive comparison to Lefebvre cannot be overlooked:

> Housing problems only arise when housing processes, that is housing goods and services and the ways and means by which they are provided, cease to be the vehicles for the fulfilment of their users' lives and hopes. [...] To be of any positive and constructive use, housing problems must be restated in terms that indicate burdens or barriers created by housing procedures, goods and services; or in terms of waste resulting from the failure to use available resources, or the misuse and non-use of resources.
> (Turner, 1976, p. 64)

Turner's experiences in Lima in Peru set about a process of analysis and contestation that would confront and briefly popularise the informal urban situations of Latin America, evidenced by the full authorship given to Turner of an entire RIBA journal in 1974. In the context of the 1970s Turner's engagement in housing as an informal process offered considerable value as an exemplar of alternative spatial practice. Yet this moment in the spotlight was short-lived, and truly crushed by the rise of neoliberal economics in the 1980s. Thus, it is only now, some fifty years later, that comparisons such as ours with Lefebvre's spatial discourse can begin to re-read and re-value Turner's advocacy for supporting and reinforcing the social relations of informal settlements and his advocacy for the perception of housing as a social process. Today, Turner offers a mechanism to re-imagine and re-appropriate both his own work and that of Lefebvre in a global and contemporary context.

It must also be noted here that our comparison does not seek to propose a simple paradox of wealth and poverty, developed or developing, or even the quantifiable compared to the qualitative. Instead Turner's discourse simply offers a concrete realisation of an architecture judged upon what it does socially and economically as a process, not what it is as an aesthetic object or product.

In an effort to ground Turner's distant housing processes in a more familiar context, here we might briefly mention the history of anarchist housing in the UK, and in particular the observations in Colin Ward and Dennis Hardy's 'Plotlanders' (1972). Ward and Hardy documented a history of alternative housing in the UK that seemed to die out concurrently with both the advent of centralised planning policies to police 'unauthorised' housing, and the economic model of increasing land and housing prices as an underlying premise of growth in the UK. It is remarkable that the self same social conditions and relationsips documented by Turner – security, acess, tenure – are again prominent in informal anarchist housing practices found right in the heart of the English countryside – perhaps the epitome of Westernised space. Yet, having emerged organically in the late 1800s and early 1900s these spaces of marginality intersected with social identities of difference and otherness to generate the last moments of documented widespread informality in the UK (Bower, R. 2016).

Today almost all plotlander sites have been lost – either to forced planning demolitions or the slow gentrification of time. Yet young Western architects are trying to remember and re-invent new ideas of anarchistic and innovative housing processes that echo much of the plotlander spirit, and Turner's advocacy for housing as a verb. Our previous reference to the WikiHouse project led by Architecture 00, combined with a revival of a plotlander model of land re-appropriation (maybe a C5 new land use planning policy most notably being advocated by Alistair Parvin?) would provide a very simple starting point from which to imagine a new sustainable housing model in the UK, built upon the underlying premise surmised by Turner forty years ago in Latin America: housing is a verb; a social process; a material dialectic.

When framed against Turner's articulation of housing as a verb and for the positive social value of informal housing, our comparison with the spatial critiques of Lefebvre's dialectical materialism begins to suggest a provocative resonance. This in turn leads towards Lefebvre's overlooked spatial contextualisation of the reproduction of the social relations of production as intrinsic to understanding the contradictions of capitalism, its survival, and the inherent possibility to contest it in social relations and practices of the everyday (1976, pp. 42, 56, 59).

Social relations of production

Whilst Lefebvre's critical re-appropriation of dialectical materialism informs the theoretical foundation of our comparisons, his later text *The Survival of Capitalism* is perhaps more important in supporting the overall trajectory of this book and the crucial appropriation of his concept of the social and relational productions of space (Lefebvre, 2009, pp. 187–189). In this focused examination of the relations of production to capitalism Lefebvre articulates a spatial appropriation of Marx's critique of the modes of production (Shields, 1999, p. 122). In contrast to institutional Marxist interpretations of the contradictions of capitalism as inherently negative, Lefebvre critiques the assumed linear causality between the social relations of production and capitalist politics of space (Elden, 2008, p. 88; Lefebvre, 1976, pp. 19–21) and generates a provocative advocacy for an alternative proposition of the positive opportunities for social change and *mondialisation* within capitalist space (Lefebvre, 1976, p. 126).

In search of an articulation of the spatial relations of production as a 'process, with a direction' (ibid.) Lefebvre applies the concept of a continuously reproducing, cyclical, and materialist dialectic to observations of the social relations of production. Echoing Marx he realised that if these relations were understood as part of the praxis and synthesis of materialist conditions, then they must be being produced and reproduced in space (Lefebvre, 1976, p. 29; Turner, 1976, p. 26). More significantly if they were being produced then they could not be predetermined or fixed (Brenner, 1997; Lefebvre, 2009, pp. 193–194, 2003, pp. 175–177). And if they were not fixed, then formal capitalist social relations of production were not a global inevitability.[19] Here this theoretical turn suggests an opportunity for a connection and critical comparison with Turner's articulation of housing as a verb as a counter to conventional hierarchical and institutionalised architecture and planning.

Lefebvre's socio-spatial and dialectic re-interpretation of capitalist space and production suggests that continued fruitless attempts to somehow defeat an imagined leviathan foe of capitalist economics head-on through direct political opposition were always destined to fail (Elden, 2004, pp. 180–182; Kristin, 1988, pp. 8–9). Capitalism is itself only a part of the social process of producing social relations. It is dynamic, adaptive, and coercive – something that Lefebvre suggests Marxism was perhaps never quite able to grasp (1976, p. 8). For Lefebvre this proposition suggested something decisive – that the coercive power of capitalist space was not held in abstract models and modes of production, but in the unconscious coercion of social relations and production of space (Kipfer et al., 2008, p. 10).

Here Turner's development practice and alternative housing models intersect with Lefebvre's proposed interpretation of social relations of production as an open and continuous socio-material dialectic (Shields, 1999, p. 158). In advocating support for the alternative spatial relations of informal settlements and facilitating their support and integration as legal and valuable city developments Turner provides a positive and practical contestation of the social relations of formal housing production. This comparison to Turner is further reinforced by Ana Paula Baltazar and Silke Kapp's work at MOM,[20] and their analysis of contemporary informality in the context of Lefebvre's social relations of production:

> He [Lefebvre] argues that the persistence of capitalist social relations is not self-evident. It is neither 'natural' nor 'obvious' that a mode of production to which crisis is inherent, manages to maintain productive forces constantly subordinated to contradictory relations of production. [...] Therefore, Lefebvre asks how capitalism maintains and renews itself generation after generation. His answer is that capitalism survives due to its capacity to produce space according to its own logic, and to accommodate any resistant niches into itself.
>
> (Baltazar and Kapp, 2007, p. 12)

Yet this interpretation of the social relations of production as unfixed also provides the foundation for a renewed critique of the social relations of production as a dialectical materialist process. Thus as with Turner's critique of the social value of housing as a process, it also places the agency of producing these relations at the heart of our comparative analysis and critique of space and capitalist ideology (Lefebvre, 1976, p. 61). Understanding the social relations of space as a continuous process generates a material and historical framework from within which to perceive social relations as spatio-temporal manifestations of broader political intent. Thus Lefebvre's proposition seeks to understand capitalism as a materialist dialectic (ibid.). Viewed in the context of this critical comparison, institutionalised forms of housing can be critiqued as inter-dependently linked with capitalist social relations and the assumed inevitability of ideological cohesion, homogenous values, and growth.

At a global level the material evidence of political coercion and social inequality can be observed (and was observed by Turner) as contested in the contradictions of permanence and impermanence in capitalist ideology. The informality that exists at the edges of capitalist space – in the slums, favelas, and barrios of informal settlements – stand in contrast to capitalist ideological belief in inevitability, cohesion, and endless growth.

Contradictions thus seem to become apparent more readily when instead of interrogating the form of capitalism you understand its production through the social praxis of peripheral space (Lefebvre, 1976, p. 17). The observation of the seemingly ideological contradiction of capitalism and informality is a continuation of Lefebvre's earlier work on the sociology of Marx, and an explicit observation of the logical fallacies that ideologies generate (1968b, pp. 116–120). However, more significant to our comparisons is the question of whether local or global scale and inequality affects our awareness of these contradictions.[21] Significantly Lefebvre suggests that the social, economic, and political contradictions of formal space are masked by the projection of ideological cohesion and are only made explicit at a global scale:

> One cannot show how the relations of productions are reproduced by emphasising the cohesion that is internal to capitalism. One must also and above all show how the contradictions are enlarged and intensified on a worldwide scale. The attempt of a separate 'theoretical practice' to superimpose the mode of production upon the relations of production, as coherence upon contradiction, has only one aim: to liquidate the contra-dictions and evacuate the conflicts (or at least the essential ones), by obscuring what happens to and results from these conflicts. [...] The dialectic is liquidated precisely at the moment when a fundamental interrogation is called for, concerning the relation between the coherence and cohesion on the one hand, and conflict and contradiction of the other.
>
> (Lefebvre, 1976, p. 63)

Lefebvre's suggestion is that the dialectic of cohesion and contradiction might only reveal itself in space when capitalist coherence becomes illogical. The plausibility of this analysis is revealed when it is compared to the expression of inequality and oppression implied by informal settlements in the Global South as 'transgressions' (Elden, 2004, p. 155; Lefebvre, 1991, pp. 396–397, 1976, pp. 34–35, 1968b, pp. 53–58; Merrifield, 2006, p. 54). This question of the peripheral global location of such transgression of capitalism is the same historical subject that Engels pursued in industrial Manchester before the globalisation of poverty removed these conditions from early industrialised Western space. The equivalent contemporary question suggests the logical necessity to consider people and social relations that exist in the informal peripheries and contradictions of capitalist space:

> Analysis of social space reveals that coherences (strategies and tactics, 'sub-systems') enter into conflict with each other. There are specific contradictions for example, those between the centres and peripheries [... but the] relation between the centre and periphery is not generated 'dialectically' in the course of historical time, but 'logically' and 'strategically'[...] We are not speaking of a science of space, but of a knowledge (a theory) of the production of space.
>
> (Lefebvre, 1976, p. 17)

In light of this explicit observation of the contradictions of a generated periphery / centre dichotomy our comparisons of Turner to Lefebvre have concordantly focused upon the dialogues and alternative spatial relations that can be observed in the informal housing advocacies of Turner: housing as a process and praxis of choice, autonomy, and social sustainability. This analysis thus seeks to engage with identities that are informal, alternative, and other as protagonists that remain subservient to the capitalist schema in search of a positive alternative praxis of dialectical materialism.

Contradictions and transgressions

In his critique of space and the reproduction of the social relations of production Lefebvre intersects the contradictions of capitalism with the inevitability of social transgressions (Elden, 2004, p. 144; Lefebvre, 1976, p. 35). The positive potential of spatial transgressions outlined by Lefebvre provokes a contested comparison with Turner whose autonomous and progressive housing model is notable for being implicitly founded upon an apparently anarchist engagement with political theory (Bishwapriya, 1994, pp. 16, 34; Harris, 2003, p. 348; Hodkinson, 2012, pp. 428–430). Critically the disjunction between anarchism and Marxism is transcended in our analysis by the similarities drawn in both Lefebvre and Turner to social relations of space as a process. Both Lefebvre's and Turner's analysis of contexts of periphery and transgression provoke an analytical and dynamic methodology that re-frames informal settlements as models of how to generate the dynamic spaces of vitality, difference, and inclusion (Shields, 1999, pp. 104, 213). In contrast to their assumed negativity, these social relations and transgressions feed off the contradictions of the capitalist form and produce something new and different through a continual and sustainable dialectic process, as Baltazar and Kapp describe:

> The richness of the 'favela', as an example of open process, space of difference and dynamic space, can still be clearly seen, although it is not guaranteed to last in a near future. We are not proposing we all should move to 'favelas' or start living without any planning. Our analysis of the 'favela' intends to indicate the formal possibilities of dynamic and not entirely predictable spaces, which indeed accommodate differences.
>
> (Baltazar and Kapp, 2007, p. 1)

The proposition therefore becomes how to learn from informal settlements and to engage with how communities can produce social relations of production and space themselves, and spatial practices that can accommodate, promote, and celebrate difference.[22] Turner's observations of informal settlements in Peru suggest a methodology or framework with which to positively engage with alternative spaces. If housing can be positively re-appraised and re-appropriated as a verb, then what can be learnt from the social and economic opportunities of informal and un-planned space, and how can we use this to engage with the new potential of alternative socio-spatial agency and grass-roots progressive development?

The point of comparison with the contradictions and transgressions affords the opportunity to connect Turner's engagement with informal housing with the spatial and urban criticisms of centre vs periphery as an economic and political construction (Lefebvre, 2009, p. 189, 1996, pp. 169–170, 1976, p. 17). This critique of 'the right to the city' and 'the right to difference' continues to pervade contemporary urban theory and will be discussed in more detail in the next chapter (Cruz, 2012; Fraser, 2012a, 2012b; Harvey, 2012; Merrifield, 2013a; Roy, 2011). However, the comparison revealed here seeks to highlight the provocative intersection of informal settlements – favelas, barrios, slums – against Lefebvre's articulation of social transgressions as inevitable expressions of difference and the contradictions of capitalism:

> This dialectised, conflictive space is where the reproduction of the relations of production is achieved. It is this space that produces reproduction, by introducing into it its multiple contradiction, whether or not these latter have sprung from historical time. Capitalism took over the historical town through a vast process, turning it into fragments and creating a social space for itself to occupy. But its material base remained the enterprise and the technical division of labour in the enterprise. The result has been a vast displacement of contradictions, requiring a detailed comparative analysis.
>
> (Lefebvre, 1976, p. 19)

This centre–periphery dialogue in itself succinctly reflects a key spatial implication of the contradictions of capitalism and social enterprise (Lefebvre,

2009, pp. 175–176). Lefebvre explicitly references 'so-called underdeveloped countries' in his articulation of the differences expressed in transgressions against the contradictions of capitalist space (1976, p. 116). Such contradictions are articulations of the exclusion and coercion of difference from accepted structural centrality of state government and political process (Mangin, 1967, pp. 69–71).[23] Yet these transgressions can also come to be identified as critical counter-narratives of the formality and structural rules and expectations of modern Westernised city models (Lefebvre, 1991, p. 373).

Here Lefebvre's positive advocacy for the appropriation and transgressions of urban space can once again be critically compared to the earlier explication of Turner's 'housing as a verb'. The material dialectic of user-choice and self-help housing is interdependent with the contradictions and transgression of capitalism, and thus the conflict between formal and informal, marginalised and accepted, central and peripheral (Lefebvre, 2009, p. 145). The urban transgressions of informal settlements and housing expressed at a global level reflect the inherent inability of capitalism to absorb and manifest a sustainable material reality and the inherent inequality of neoliberal economics.[24] Thus the identities of transgression and illegality against socially accepted patterns can be interpreted as a reaction to the material reality of inequality.

Turner realised that the development of informal settlements he documented were in fact logical and reasoned actions of people generating rational answers to their situation through the illegal inhabitation and production of space. This process and identification of informality and urbanisation became a performance between the police and squatters that Bromley describes as 'an elaborate charade' (2003, p. 274). This reality has been somewhat successfully suppressed and hidden from cities and space in the Global North, however its global prevalence remains a depiction of a global ideology of the inevitability of continuous growth and a rejection of the finite reality of global resources and economy (Bauman, 2000, pp. 36–37; Chang and Grabel, 2014, p. 25; Sachs, 2010). Thus Lefebvre's identification of the positive potential of difference as transgressions against ideological cohesion is supported by Turner's progressive, intergenerational, and sustainable facilitation of informal housing practices (Turner, 1976, p. 62).

Examples of alternative and 'sites and services' do exist in the Global North. Perhaps the most notable contemporary example of alternative housing has emerged in the Netherlands in the twenty-first century, and particularly the city of Almere.[25] The development model in Almere is built upon a complex and layered masterplan by Rem Koolhaas and OMA. Yet the key underlying housing principle of the masterplan is a contemporary Western equivalent of the 'sites and services' model. It offers plots of land at a subsidised rate of 375 Euros per m2, and then allows individual and group housing developments almost complete planning autonomy in terms of the type and style of house that emerges on each site (though crucially maintaining strict environmental requirements). The resulting places and community are so recent as to still resist any definitive analysis of either a positive or negative outcome, yet the mere existence and positive uptake of a project like this is itself a sign that everyday people do desire alternative visions of housing (Hall, 2013, pp. 154–162). And in particular (and perhaps unsurprisingly) it seems that it is the alternative economic model that is perhaps the most compelling part of the opportunities of alternative housing.

Both the visual and social alterity of projects like this seem more explicit against the background of conventional Westernised space. Almere stands as a long-term experiment that questions whether making land affordable through subsidy will have long-term trickle-up economic (and theoretically social) benefits. Yet it also stands as a representation of contradictions and transgressions against normality, sparking derision, confusion, and even contempt. Anecdotal evidence and comments from online architecture journals can be noted as reflecting a dismissal of the strange, weird, and wonderful architecture produced by home builders at Almere. As with the spontaenous housing of plotlanders, the challenge to taste in such architecture is balanced by the challenge to the professional authority of architects to define what is acceptable as an architecture language of housing. Whilst on the surface this appears a valuable asset to the professional identity it also suggests the limitation of our aspirations to engage with the social relations of space at merely the level of good taste, relinquishing the need and desire and responsibility to engage on social and political levels.

Even momentary fascination with such projects in the Western architectural media only seems to reinforce the lack of traction, uptake, or even experimentation with positive housing alternatives. Thus, for example the neoliberal politics and economics in the UK lead to the 2014 announcement of 'new garden cities' that pander to conventional development models and capitalist restrictions of commercial viability. Due to political (and social) hegemony, it is almost a fait accompli that once again we will unfortunately end up with houses and new cities that exist as objects of capital – nouns – instead of real places of social relations and process – verbs. The question of material or social necessity appears to remain unchallenged, and cultural hegemony of inevitability prevails.

Necessity, informality, periphery

Due to preconceptions of alternative and informal space as a conflict against political normality and formalism, informal settlements remain largely isolated in social negativity. Thus large areas of the world – the ambiguous 'Global South' – are dominated by apparently negative and transgressive space. The existence of informal settlements is deemed symptomatic of a violently rapid urbanisation of huge populations and the inevitable inability of formal city structures and political systems to adapt to this pressure and to provide access to these necessary social and economic networks (Bromley, 2003, p. 4; Neuwirth, 2006; Roy, 2011, 2005). Such settlement practices are driven by the well established economic, social, and cultural processes through which rural populations migrate to rapidly urbanising cities and proceed through staged and layered processes of integration into social and economic networks (Mangin, 1967, p. 68; Pugh, 2000). Yet accounts of informal settlement development make clear that various levels of economic stability are manifest within these communities as part of their social and spatial development. Once again, Baltazar and Kapp succinctly describe the Brazilian expression of these issues:

> Some of the big Brazilian cities, such as Belo Horizonte, are just over 100 years old. When this city was 'founded' (it was a designed city) it offered a place for an elite to live in accompanied by their workers. As the city grew, there was a need for more workers along with the many informal activities which started taking place. This growth was not planned, and since the model of the city was very rigid – there is even a contour avenue supposed to fix its spatial limit – it was not prepared to accommodate the ones who were not programmed to be there. It is a model of exclusion imposed by spatial design. [...] 'Favelas' are born in response to this rigid and exclusive city model, in order to accommodate those workers and those looking for work in the new growing city.
>
> (Baltazar and Kapp, 2007, p. 1)

These same observations of necessity and contradiction are at the core of John Turner's much earlier experiences of Peru in the 1960s as he encountered the implications of informal settlements that were beginning to take root and expand in the surrounding urban periphery of Lima (1974). The speed and dynamic adaptation of informal settlements, coupled with the necessity of urbanisation generated a social and spatial methodology that is intrinsically a material expression of necessity and informality. In contrast to the centrality and hierarchy of structural space and state housing, the social relations produced by informal settlements cannot be reduced to abstractions and objects, existing as they do within distributed and localised socio-economic networks (Turner, 1972a, p. 152). Viewing informality as a contradiction informs a political isola-tion of their interdependent alternative social relations as counter and negative appropriations of space. Their ability to produce novel and dynamic social rela-tions in reaction to the capitalist contradictions highlights the socio-political and spatial isolation that Turner encountered. Yet informal space remains de-valued and unable to transcend this negativity (Baltazar et al., 2008, pp. 12–13).

At this point it should be made explicit that Turner does not romanticise informal settlements. His observations are not an attraction towards some fantasy of impoverished utopia, but a stark reflection of inequality that was only beginning to be realised in the 1960s. Yet the global prevalence of informal settlements and urban inequality today allows Baltazar, Kapp, and Morado to provide an appropriate summation of a conflicting positive and yet harsh reality:

> An everyday production of space, which in some aspects resembles the idea of emancipation, happens in Brazilian favelas today. Nevertheless,

the favela space should not be romanticised as it occurs out of necessity not choice. The relative autonomy of the favela dwellers in the production of their spaces is a direct consequence of their marginal position in the economic system, which excludes them from the consumption of architecture as a formally produced commodity. Any of its possible advantages are born out of its antagonisms within the socially dominant order.

(Baltazar et al., 2008, p. 18)

As observed and documented by Turner, the existence of informal settlements is in fact merely a highly appropriate material and spatial resolution of the political and economic context in which people are having to live (1968, pp. 356–357). It was (and remains) in essence a logical process of dialectical materialism. A material response and dialectic process of necessity and survival practised non-hierarchically at grass-roots level. Significantly this comparison suggests that Turner's advocacies reinforce social relations that generate something more than the apparently crude and insubstantial dwellings. The process of people generating their own settlements outside formal authority allows them to create, utilise, and continually re-create networks of social relations that directly improve and support the identity, stability, and prosperity of individuals and communities. Writing fifty years later, Fernández-Maldonado identifies the significance of the strategies of engagement with material and social inequality and necessity as the key element of study that precipitated the unique research generated by Turner and his contemporaries (2007, p. 5).

Conversely, Dan Hill's work with SITRA and housing in Helsinki is a valuable Western and contemporary example of an engagement with these issues and Western interpretations of necessity. The 'Low2No' housing project (design competition completed in 2009) was a proposal to deliver a housing development in Helsinki that would challenge a number of assumptions about conventional housing models – timber building standards, the relationship of local food and restaurants to commercial developments, and the economic viability of alternative commercial models. As Hill describes it, the project itself was a vehicle – a McGuffin or Trojan horse – that allowed SITRA to engage with and question larger issues of the social relations of production (2015, p. 104).

Yet is important to recognise that this work all existed under a framework of government backed strategic design research, and in spite of this support, still suffered from an inability to see ideas translated into reality for an extremely protracted planning and development period, only being realised on site in 2015. The legacy of the work exists in the strategic documentation and research that supported the projects rather than in physical manifestations. Westernised space and socio-spatial relations do entail a seemingly intransigent context in which to enact alternative approaches to space, and this is all the more explicit when considered in comparison to the apparently primitive and backwards contexts of informal cities and spaces in the Global South. Thus we return to Turner to observe the positive potential and implications of actually realising actual change to the social relations and production of space.

By validating an alternative way of producing space that worked with the existing creativity and innovative dynamic people he worked with in Latin America, Turner helped to reinforce the social production of alternative relations of production that would contest formal political and urban values. It created 'a process which was vividly described in "Desborde popular y crisis del Estado" (Popular overflow and crisis of the State) by Matos Mar (1984) who claimed that these new practices were altering the conventional social, political, economic and cultural "rules of the game"' (Fernández-Maldonado, 2007, p. 5).

The inherent fear in the formal identification of a 'popular overflow and crisis of the state' is a direct response to the ability of a vast and impoverished working class to 'alter the conventional rules of the game' (Bishwapriya, 1994, p. 34; Rapoport, 1987) at social, cultural, economic, and even political levels. Notice that the threat identified in changing the rules of the game is not aimed at a supposed illegality of the settlements, but at their social impacts (Baltazar and Kapp, 2007, p. 8; Pugh, 2000, p. 332). It questions how such spaces and relations challenge the urban condition through the creation of associative practices, enterprises, business, and so on, or in essence, the production of their own social relations of production and space. As Baltazar and Kapp suggest, these practices are in direct opposition to the assumed social passivity and subordination of informality:

> 'Favela' is then an answer of a modern spatial attempt of inclusion, focusing on difference and the dynamic possibility of growth in order to accommodate the ones that are excluded from the planned city. Although the reason of existence of 'favela' is related to the need to 'solve' a spatial problem, its developments are strongly committed to the problem-worrying strategy.
>
> (Baltazar and Kapp, 2007, p. 1)

The social and visual discomfort directed towards informal settlements from the Western perspective can be understood as merely evidence of anxiety at the alternative social identities and practices produced by those succeeding and prospering within informality (Pugh, 2000, pp. 332–333; Roy, 2011, p. 232). This is an uncomfortable inversion of the assumed passivity of those who were deemed excluded, isolated, and peripheral. Having placed so much stock value (both figuratively and literally) in the unquestionable supremacy of the formal housing and socio-political processes and institutions, the expression of something so evidently counter to formal and regularised capitalist relations of production is cause for political concern and socio-economic discomfort (Lefebvre, 2009, pp. 274, 301; Ward, 2008, p. 305).

This social and political discomfort also has to be measured against the realisation that these settlements are not a form of direct opposition. Instead, and as was proposed by Lefebvre, they are merely an expression of the contradictions within capitalism. They reflect an expression of the same process of dialectical adaptation without the top-down rigidity of form and hence producing social relations of inclusion and economically realistic sustainability (Neuwirth, 2006, pp. 62–65). As observations of these affects, Baltazar and Kapp distinguish two key factors to the social relations of informal settlements:

> As such, the purpose of a 'favela' is not free from the system of dominance; on the contrary, it is created in order to enlarge the space of inclusiveness of the city. With regards to its formal manifestation, it ends up as an unprecedented artificial settlement inside the modern tradition. It is a dynamic space; it is alive, spontaneous, constantly growing, constantly in transformation. It is formally non-representational although it is created in order to achieve the patterns of living in the city. Its formality is a consequence of a non-planned, non-rational settlement, giving place to a more sensible manifestation, even if not intended, since it lacks predictions. The difference of the lack of prediction in 'favelas' and the lack of prediction in the city is that in the first it results from a dynamic and inclusive space while in the second it is a consequence of an exclusive plan ending up as a static and exclusive space.
>
> (Baltazar and Kapp, 2007, p. 1)

Lefebvre's proposition that contradictions are only made apparent at a global scale focuses attention onto the geographical, socio-economic, and political peripheries as the arena in which the potential for alternative social relations of production might exist. This subsequently reveals places that might provide the opportunity to produce spatial relations different from 'any that can be inferred from the existing relations of production. [... relations] produced through space as well as time, and by means of a conception of space' (1976, p. 35). Lefebvre realised that capitalism's power wasn't manifest in any fixed idea of production or abstract inequality, but in the process of consumption itself. Thus, in comparison Turner's alternative advocacy for the social production of informal housing once again marks a uniquely practical and positive advocacy that resonates profoundly with Lefebvre's articulation of transgressions and difference to produce alternative and sustainable social relations of change.

Critique of housing as a verb

Turner was not alone in the 1960s and 1970s in his questioning of the implications of projecting Western models, particularly in the context of development as a global ideology affecting Latin America and the wider Global South.[26] Colin Ward notes similar critical reflections being made by architectural contemporaries such as Giancarlo de Carlo, as well as in the political discourses of Ivan Illich (1976) and Paulo Friere (1996). In the context of Turner's discourse both Illich and Friere provide useful interrogations of the social implications of formal spatial development models and observations of the interdependent relationship of development and social identity (Ward, 1972, p. 4). These intersecting interpretations resonate with Turner's observations of the mismatches of state based housing, and still pervade the contemporary conflict of formal and informal development (Hodkinson, 2012). The methodologies and practices that produce space and communities are inherently connected, being both subject to authoritarian intervention yet also holding an inherently positive potential for change (Baltazar et al., 2008, pp. 12–13).

Further broadening the critical framework and theoretical potential of his discourse, retrospectively Turner would describe and utilise connections from his practices to the loose-fit principles of Alex Gordon (1974), Simon Nicholson's 'The Theory of Loose Parts' (1972), and John Habraken's *Supports* (1972). Yet it is equally important to highlight the contemporary criticisms that generated an overtly socio-political contestation of the implications of Turner's advocacies for autonomous housing. The main trajectory of such critiques suggests that Turner's development models implicitly allow the state to relinquish its responsibilities to its people, generating housing models of sweat equity and neoliberal co-option (Harms, 1982, 1976; Ward, 2005, 2004, 1999, 1982).

The most notable of these critiques is that of noted neo-Marxist and structuralist Rod Burgess who engaged in provocative debate about the implications of a 'Turner school of development' (Burgess, 1978a, 1978b; Jenkins et al., 2009, p. 24). His critique suggested that true choice could not be achieved by self-help housing models, which would be inevitably co-opted by systems of structural constraint, namely, poverty and the lack of effective choice. Burgess' Marxist critique also focused upon the potential de-densification implicated in self-help models, suggesting a prominent challenge to our comparisons with the explicitly urban discourse of Lefebvre (Burgess, 2001). For Burgess informal settlements could not function outside capitalism and market relations, and therefore self-help focused excessively on use-value rather than on exchange value of housing (1982). Abandoning people to their own devices and relinquishing the responsibility of state support for the individual and collective principles of society ran counter to the principles of Marxist-based state socialism that were and remain the core doctrine of the political and academic left.

Here it is important to note the contrast between Burgess' institutional articulation of the social revolutionary nature of Marxism as arriving through direct political struggle, and Lefebvre's engagement with the inherently positive spatial and dialectic potential of implicit difference, appropriation, and the spontaneity of urban social relations to achieve change (Elden, 2004, p. 144; Merrifield, 2006, p. 108). Instead Burgess' critique focuses on the implications that surrounded global development and the co-option of informal housing (Burgess et al., 1997b, p. 147). Robert Harris would seem to clarify these contradictions in his highlighting the misrepresentation and simplification of

Turner's discourse to a programmatic model of sites-and-services as a panacea that Turner notably never sought to provide (2003, pp. 260–263). The later political adoption of 'self-help', 'sweat equity', and progressive housing models by organisations such as the UN and World Bank coincided with global economic models of neoliberalism, leading to the adoption and co-option of Turner's ideas. Removed from their practical contexts and stripped of their political intentions, the remaining principles of economic efficiency resonated with 1980s neoliberal economics whilst fundamentally disintegrating the underlying premise of user-defined housing and social development (Turner, 1992). As Harris identifies, the most innovative contributions Turner made in advocating the 'political necessity of user choice' are largely overlooked (2003, pp. 263–264). Thus, Colin Ward notes:

> Notice that he says 'design construction or management'. He is not implying, as critics sometimes suggest, that the poor of the world should become do-it-yourself house builders, though of course in practice they often have to be. He is implying that they should be in control.
>
> (Ward, 1972, p. 6)

In contrast to mere 'sweat equity', Turner's proposition is a far more fundamental political contestation of authority and value, articulated through a simple and practical analysis of housing. Thus he notes that the most important thing about housing is what it does in people's lives, or in other words that: 'dweller satisfaction is not necessarily related to the imposition of standards' (1992, p. 5). This premise is reinforced by the contestation of value implied in his observation that 'the deficiencies and imperfections in your housing are infinitely more tolerable if they are your responsibility than if they are somebody else's.'

Thus, within the demonstrable economic and socially logical principles of progressive development Turner was evidently aware of the implications of the social content and relations that this process was generating in relation to concepts of autonomy, freedom, and so on (1968, pp. 357–358). Yet crucially, and in contradiction to Burgess' critique of self-help as a project, Turner had not imposed these practices as an external political influence upon the context of informal settlements (1976, pp. 37–40). This was not an alternative economic, political, or state-imposed ideology. It was not the imposition of a socialist, Marxist, or even anarchist political ideology. Instead, Turner was observing, documenting, and eventually facilitating social relations and practices that were already occurring. Having simply found a culture of self-build housing autonomy, and recognised it as a logical expression of social and economic contradictions, his opportunity to observe and interact with the reality of people's everyday creativity and entrepreneurship allowed Turner to document what remains a valid concrete expression of the positive potential of autonomy. This interdependence of autonomy and informal settlements continues to be highlighted in contemporary contexts, as exemplified here by Baltazar and Kapp:

Autonomy in the design or production of space means that people involved in designing and building need to have access to knowledge of design and building processes and components in order to discern and enact. But at the same time it means that those processes have to be open enough to increase autonomy instead of limiting it or even turning it impossible.

(Baltazar and Kapp, 2007, p. 10)

As previously discussed, the apparent socio-political opportunities that are created in spaces of marginality and exclusion need not be interpreted as any form of Marxist or socialist utopia that might promote an abstract alternative or provide anything remotely approaching an ideological polemic (Baltazar and Kapp, 2007, p. 8; Lefebvre, 2009, pp. 100–106). Any attempts to do so would be counter to Turner's original critique that diligently pursued practicable and materially grounded responses to the contradictions of informal contexts and economies of absence. Yet crucially, his methodologies would appear to resonate with aspects of the socio-spatial discourse of Lefebvre; housing and development re-imagined as a process of generating sustainable alternative and positive social relations (Turner, 1997, p. 164).

Thus our premise remains that informal settlements can be re-read and compared as concrete spatial realisations of Lefebvre's observed contradictions of capitalism, and reciprocally, that Lefebvre's apparent Marxist principles of social space are perhaps most plausible and realisable in the context of informal and anarchistic models of development practice (Firth, 2012, pp. 81–84). Such a re-reading reinforces the observation that Harris' and Burgess' criticisms are explicitly not aimed at Turner, but at the narrow political appropriations of his work. By transcending such a narrow reading of Turner's work this book re-aligns his advocacy for user choice and autonomy as a potentially revealing realisation of Lefebvre's socio-spatial and materialist framework of critical discourse. It looks to appropriate the positive socio-economic potential of positive plural choice progressive and participatory housing, whilst understanding that sweat-equity is merely a practical reality of economies of absence, not a solution, ideology, or utopian vision.

If the spatial and social relations of informal settlements are simply the logical response to intrinsic contradictions of capitalism represented at a global scale in all its inequalities, then Turner's premise of user choice and autonomy can be read as a re-valuing of these spaces as positive global articulations of social difference and transgression that might be re-appropriated as active political contestations. Thus perhaps of greater concern than the delineation of Marxist and anarchist concepts in his work is the continued lack of political engagement and recognition of informal settlements for what they are, and the continued perception and uncritical interpretation of informality simply as a reaction to the periphery's refusal and structural inability to form a logical cohesion (Baltazar and Kapp, 2007, p. 3).

Here the question of access to 'political articulation' becomes both a validation of Turner's overtly political engagement with development (Turner, 1976, pp. 155–162), and a challenge to the potential of informal settlements to become articulated beyond their current identity of exclusion and periphery. In the context of both Turner's practices and contemporary conditions in the Global South, the demonstrated socio-cultural beneficial value to communities existing outside formal control is offset against their intractable lack of advocacy, interaction, and voice at a political level (Burgess et al., 1997a, pp. 150–152). This affords our analysis a crucial renewed intersection with Lefebvre in the critical comparisons of spatial autonomy. In this context Lefebvre notes that whilst the global phenomena of informal urban spaces and settlements exist, they remain socio-culturally, politically, and semantically excluded as a periphery. The potential value of such informal, alternative, and different spaces remain isolated and cannot achieve their true potential to contest the existing social relations of capitalism:

> This tactic of concentrating on the peripheries is not wrong, in fact the very existence of the peripheries is symptomatic of the importance of the 'centrality' which operates. [...] The masks and snares of power are revealed in their full light, and the ideological clouds are dispersed. [...] And yet this tactic, which concentrates on the peripheries and only on the peripheries, simply ends up with a lot of pin-prick operations which are separated from each other in time and space. It neglects the centres and the centrality; it neglects the global.
>
> (Lefebvre, 1976, p. 116)

Autonomy and heteronomy

We have observed that Turner's advocacy for informal settlements can be positively compared against the negative implications of ideological political and economic constraints of formal urban models (see: 'Cultural Values and the Economy of Autonomy', in Turner, 1974). In reaction to the economic inefficiency of formal development practices Turner recognised the opportunity and necessity to facilitate the removal of objects and barriers that restricted the progress of these communities and advocate for them at an economic and political level (Sandercock, 1998, p. 37; Turner, 1967, pp. 177–179):

> In other words, to state the problem of housing (or any other personal and necessary local service) depends on who needs the statement and what it is used for. If housing is treated as a mass-produced consumer product, human use values must be substituted for material values. [...] However sensitive individuals in such heteronomous systems may be, they are locked into positions in which this contradiction is inescapable.
>
> (Turner, 1976, p. 66)

Turner's involvement in various NGOs provided him the ability to advocate initiatives that would benefit and strengthen socio-spatial relations that crucially already existed in informal settlements. This simultaneous act of valuing and advocating the positive potential of informal space allowed Turner to promote spatial methodologies with which to empower communities with the authority that comes from social choice and change (Bishwapriya, 1994, pp. 34–35). Such spatial and political initiatives ranged from financial loans to home-owners, to increased availability of building materials and advocacy for rights to ownership of land to stabilise tenure. Yet all of these examples were intrinsically linked to reinforcing the network of user choices that autonomy was predicated upon:

> If housing is based on open services, the builder, buyer, or house-holder is free to combine the discrete services in any way his own resources and the norms governing their use allow. In other words, local executive decisions (and generally supra-local normative decisions) are fully differentiated. For the local decision-maker or user, the open service system has a high degree of, or the capability for, providing many different ways of achieving the same end – in the present case, the construction of a house.
>
> (Turner, 1972a, p. 154)

Here, Turner's simplistic explanation of autonomous and network based relations of production nevertheless provided a clear expression of why top-down interventions were an inappropriate, restrictive, and homogenous response compared to informal settlements (1996, p. 344). At a time when modernist housing super-blocks were widely utilised to re-house people who had been forcefully evicted from informal settlements (Kotanyi and Vaneigem, 1961) Turner's analysis was sorely needed (1976, pp. 46–47; Ward, 2008, p. 296). However, whilst valuing the vast increase in choice offered by heteronomous housing procurement models, Turner also recognised that 'expert systems' remained explicitly necessary as mechanisms to facilitate and support existing community frameworks (1996, p. 345). Such positive 'expert systems' included the necessity to support local builders with structural and safety expertise, to plan efficient typology patterns to guide and inform those who ask for help, and to engage with and support communities through grass-roots action and participation. Yet beyond these elements of practical support, for Turner it was the necessity to advocate with state political bodies for improved amenities and legal rights and to resist any centralised planning that was ultimately the key aspect of his housing development methodologies (Franks and Turner, 1995).

The autonomy Turner defined in the simplistic contestation of 'who decides, and who provides?' (1976, p. 127) offered the basis of an alternative model and contestation of spatial and political authority and control. Thus the issue of owner-builder is not important (Harris, 1998, p. 248). For Turner, '[t]he best results are obtained by the user who is in full control of the design, construction, and management of his own home,' but, 'it is of secondary importance whether or not he builds it with his own hands' (1976, p. 158). The question of how to define what the 'best results' of housing might be explicitly tackles the issue of who decides what are the right values that our built environment engages with and embodies. This simple advocacy is that increased autonomy and heteronomy in development practices and building programmes leads to housing and communities that are best able to suit the changing needs and circumstances of their occupants. Thus the broader positive social implications of informal space and housing provide the qualification for governments and communities to engage with the autonomy and heteronomic processes of housing as a verb.

For Turner homes that are built, managed, and adapted by the occupants and owners themselves reflect a Lefebvrian realisation of the social production of space. The principles for autonomy and heteronomy in housing once again provide compelling comparison to Lefebvre's critique of the relational production of space and social relations of reproduction. The issues of autonomy and heteronomy intersect with the transgressions and differences of formal capitalism that 'endure or arise on the margins of the homogenised realm, either in the form of resistances or in the form of externalities (lateral hetero-topical, meteorological)' (Lefebvre, 1991, p. 373). Rather than existing as finished objects of consumption, informal space and housing remain dynamic and adaptive to suit the needs of the inhabitants within the reality of an economy of absence. Thus the rationality and sustainability of informal settlements suggests that they can be considered as adaptive and successful due to exactly the same methodology as capitalism (i.e. the production of social relations of production and space), but do so within the contradictions of inequality that capitalism seeks to repress and deny. The significance of the reality of inequality within informal settlements and its negation of ideology through rational materialism defines Turner's discourse:

> [If] housing is treated as a verbal entity rather than as a manufactured and packaged product, decision-making power must, of necessity, remain in the hands of the users themselves. I will go beyond to suggest that the ideal we should strive for is a model which conceives of housing as an activity in which the users – as a matter of economic, social, and psychological common sense – are the principal actors.
>
> (Turner, 1972a, p. 154)

Intrinsic to this proposition is the critique of the political and spatial practices of top-down systems of government and housing. Instead of this, Turner's insights suggest an alternative where governments need only respond to the quantitative information that points towards pent-up demands and needs by providing the materials, finance, and opportunities for people to create their own solutions (1976, pp. 56, 72). In the 1970s and 1980s this reframing of the question of urban squatter housing led to widespread political critique and laid the conceptual groundwork for criticism of the state from both the right and left of the ideological spectrum and the contrasting proposition of good governance and the leveraging of social capital. It was only later that international agencies begin to begrudgingly acknowledge the limits of self-help and the government's role in the delivery of goods and services, including housing for the urban poor (World Bank, 1997; Sanyal et al., 2008, p. 16). Yet in spite of the political and academic discourse that emerged from this period the same mis-matches and ideologies of housing are still observed as pervading urban thinking throughout the Global South, where the ideological image of development remains constrained to the vision offered by Westernised formal space (Merrifield, 2013b, p. 79).

Ultimately, Turner's observation, definition, and advocacy for progressive development demonstrated that a grass-roots, bottom-up, and networked approach was systematically a more materially and socially appropriate means to contest inequality and poverty on a global scale (1967, pp. 177–179). Progressive development practices demonstrated not only that they generate more economically and socially appropriate spatial forms, but more significantly, how the process itself generates something more. In contrast to the Marxist critiques of Burgess,[27] Turner's observations of the network of social interconnections that autonomous progressive development created resonates with Lefebvre's concepts of autogestion. He understood and believed that this process did not impede social mobility, or trap people in poverty, but actually empowered them and their community with diverse opportunities to produce alternative informal social relations and sustainable opportunities for growth:

> The significance of the cultural change that takes place over time and in the same barriada location not only confirms this kind of dwelling environment as a vehicle for social and economic development, but also points to the connections between the different demands of various social levels. It is clear that the relative priorities and demands of the low-wage earner and that of the high-wage [...] earner must be different though not as different as the levels compared above. Preoccupation with material status is as evident in the barriadas as it is elsewhere.
>
> (Turner, 1967, p. 179)

The distinct difference between politically expedient top-down practices and the alternative progressive development that Turner advocated corresponds succinctly with Lefebvre's discourse on autogestion and community self-management (2009, pp. 148–149). This convergence with Turner's advocacy for facilitating autonomous networks of social relations to generate heteronomous housing choices reflects the concrete observations of practices that existed within the social contradictions of capitalism without his prior intervention. The contradictions between the hierarchy and authority of formal dominance and the rich autonomy of informal and progressive housing marks a crucial practical contestation of Lefebvre's theoretical autogestion:

> This dominant order means, first of all, heteronomy or that individuals and primary groups are no longer able to negotiate and to decide for themselves. Even if participation is part of public policy, the whole

process of the production of space turns out to be bureaucratic, far from the understanding of most people, and dominated by so-called 'technical' decisions. Therefore, one of the main goals of a critique is to show how the general and abstract logic of the production of space determines people's lives and forces them into a passive role.

(Baltazar et al., 2008, p. 12)

Retrospectively Turner would himself come to reflect concerns and reservations of the impact and implications of self-help and sites and services (Turner, 1992, p. 6, 1986, pp. 14–16, 1967, p. 190). Yet in spite of these retrospective reflections, re-reading Turner's practices against the concept of autogestion still reveals a valuable opportunity to perceive self-management of housing in 1960s Peru as a practical realisation of Lefebvre's spatially contextualised autogestion. In this we can positively contest a renewed re-examination of Turner's practices as advocating community and social engagement with the politics of freedom and choice to take control of housing from a grass-roots level.

Autogestion and self-management

In his appropriation of concepts like autogestion and self-management Lefebvre sought to provide a further socio-spatial extension of Marxism. The term autogestion literally means self-management, but Neil Brenner and Stuart Elden note that its French connotation may be captured more accurately as 'workers' control' (Brenner and Elden, 2009, p. 14). Here Lefebvre's Marxist interpretation of workers and control can be brought into close comparison with Turner's anarchist housing premises of progressive development and user choice (Lefebvre, 2009, pp. 139–141). Thus for Lefebvre:

The aim is to take over development, to orient growth (recognised and controlled as such) towards social needs. Whoever talks about the self-determination of the working class or about autonomy, is also talking about self management.

(Lefebvre, 1976, p. 40)

This concept and practice of self-management provides an original response to the Marxist problem of how to socialise the means of production. Lefebvre notes that autogestion as a 'concept and practice can avoid the difficulties which, since Marx, have arisen in the experiment with authoritarian centralised planning' (1976, p. 120). Here a comparison of autogestion with Turner's principles of progressive development as a social practice offers a clear contestation of the same authoritarian centralisation of authority and control. Yet as with Turner, Lefebvre is explicit that self-management is not a panacea, as it poses just as many problems as it does potential solutions (1969, p. 84).

Thus autogestion as a social principle of grass-roots political self-governance is a concept that has to be fleshed out and contextualised across the full spectrum of global conditions (Lefebvre, 2009, pp. 193–194). It is in this process that Lefebvre maintains that class and workers' struggle can be stimulated through social participation, and that such active engagement in space is necessary to give self-management continued meaning. The inherent spatial practice of development and its articulation of continual social relations of production are required to resist the manipulation and potential ideology of political co-option (Lefebvre, 1969, p. 68). Thus Lefebvre makes clear his belief that only through self-management and the continuous dialectic contestation of social relations can participation be considered real (1976, p. 120).

This intersection of participation is further reinforced by the similarity in both Lefebvre's and Turner's discourse of grass-roots control and self-determination (Lefebvre, 2009, p. 150). Lefebvre's suggestion that networked and territorial autogestion should be articulated to exert pressure against state powers and administrative rationality highlights the interdependence of the transformation of social life suggested by autogestion with the material reality of political and economic obstacles is what maintains its political potential (2009, p. 250). Yet whilst Lefebvre's critique resonates with Turner's discourse and advocacy, if read in abstraction from material and social context and agency it remains empty and lifeless; becoming akin once again to Hegels' abstract dialectic logic that Marx originally rebelled against:

> The worshippers of the total state economy, for example, may use the self-management thesis: but they are just playing with words. The self-management slogan cannot be isolated, for it is born spontaneously out of the void in social life which is created by the state; it has sprung up in various places as the expression of a fundamental social need. It implies an overall project designed to refill the void, but only if it is made explicit. Either the social and political content of self-management is deployed and becomes strategy, or the project fails.
>
> (Lefebvre, 1976, p. 120)

This comparison of progressive development to self-management is made more compelling with Lefebvre's articulation of the inherent problematic that autogestion poses (2009, p. 16). Interdependent with autogestion as a potentially global social project, the complexity of social and political relations provides a direct connection to the material reality of contexts that cannot be abstracted.[28] Instead Lefebvre articulates autogestion as a dialectic process: 'What this determines is not a state but a process, in the course of which new problems are posed and must be solved in social practice' (1976, p. 125). Framed in this way autogestion is both a project of radical democratic governance and interdependently a conflicting and contradictory process.

Examples of the socio-economic potential of autogestion can occasionally be found in a Westernised context, existing as variations and experiment with projects of cooperative housing. Yet as we see in the limited number of contemporary examples raised in this chapter, the lack of popular recognition and engagement with alternative housing models can be attributed to the cultural hegemony of Westernised spaces and societies. This ambivalence and indifference to change appears to reflect a lack of tolerance to the potential of difference and multiplicity in development and space. Yet this overly simplistic analysis must be challenged and confronted. Thus, the challenge of cultural hegemony in Western space is a subject that we will begin to explore thoroughly in the next chapter.

In comparison to Lefebvre's positive articulation of autogestion, participants in progressive housing and self-management can be considered as engaging 'in self-criticism, debate, deliberation, conflict, and struggle; it is not a fixed condition but a level of intense political engagement and "revolutionary spontaneity" that must "continually be enacted"' (Brenner and Elden, 2009, p. 16). The positive potential of political change driven by the social practice and production of relationships and space offers an unrealised yet tantalising proposition:

> The analysis which I have attempted here points to the dissolution of the state, a kind of wavering away of its power, its strategic capacity and the ramifications of absolute politics. To this extent, the state self-destructs; the conditions in which it functions, its social 'base', are undermined, even though its foothold in the economic sphere remains firm. It is the institutions and ideologies, the superstructures upon which the absolute state is erected that crumble.
>
> (Lefebvre, 1976, p. 125)

These are the radical yet plausible implications of self-management responding to the contradictions and inequalities of capitalist space and generating rational and logical practices and a concurrent network of social relations. The comparative analysis methodology that this book employs to connect Lefebvre to Turner has been an explicit attempt to highlight in informal settlements the practical and positive examples of self-management that Lefebvre only hints at existing in the global periphery. What remains in our study therefore is to explore the wider implications and significance to contemporary spatial practice and practitioners of Lefebvre's observations of the informal periphery (Baltazar and Kapp, 2007, p. 8).

The contradictions and disparities of informal settlements validate the intersection of Turner's principles of autonomy and user choice, and Lefebvre's socio-political aspirations for autogestion. The explicit economic, social and political connections between the act of building itself and the social relations that these productions generate are clearly demonstrated in Turner's contrasting of instant and progressive development practices (Lefebvre, 1976, p. 121). Thus, the autonomy of informal settlements generates plurality and a dynamism that are typically the hallmarks of the capitalist process, yet their appearance and unruly reality confounds the conventional sanctity of logical cohesion. Such unconventionality is merely the consequence of what Turner perceived as user-defined choice and autonomy (Baltazar et al., 2008, p. 18).

Turning this observation on its head prompts an uncomfortable re-reading of Westernised space. Questions emerge of whether formal and conventional space offers the same plurality and freedom of choice found in the freedom of informality. Furthermore, whilst Lefebvre's theoretical critique of space has always contested the political ideology of Western space, Turner's analysis of top-down state programming versus network based social relations of progressive development places similarly Western spatial values and condition in sharp focus (Lefebvre, 1991, pp. 81–83, 1976, p. 68; Turner, 1976, p. 68). The apparent freedom of our choice is in fact largely prescriptive to a plethora of culturally and economically acceptable formalities that we perceive as freedom of expression (Firth, 2012, pp. 198–208; Merrifield, 2006, p. 154). This reality might suggest that perhaps our position at the pinnacle of capitalist space affords us little ability to generate the dynamism that exists on the periphery:

> It is impossible to induce or program such a process, or even to organise it in the manner of the industrial production. Nevertheless, it has certain objective conditions: first of all, the absence of domination in the relations of production. This implies, among other things, the disposal of the producer over her/his means of production.
>
> (Baltazar and Kapp, 2007, p. 7)

As exemplified by Turner's user choice and progressive housing, the capacity to self-generate organisations and social relations within the context of self-management practices is the foundation of the social value of informal settlements. As he was to observe, the ideological intervention of architects and planners within the dynamic realities of these communities and contexts lead to inevitable degeneration of the social sustainability of such dynamic communities (Lefebvre, 1976, p. 122). As we discussed previously, what is required is therefore to pursue the analysis of the periphery and ascertain the possibilities and implications within these methodologies. The globalised contradictions of capitalism are themselves an inevitable dynamic and shifting context of urbanisation, poverty, immigration, and spatial inequality. Yet the underlying inevitability of the existence of peripheral and informal space continues to provide opportunities for the generation of alternative social relations that potentially hold the key for the dissolution of ideology and the re-articulation of growth as sustainability. Lefebvre's articulation of this project is eloquently presented in the conclusion to *The Survival of Capitalism*:

> A strategy which would join up the peripheral elements with elements from the disturbed centres. [...] An operation of growth towards specifically social needs and no longer orientated towards individual needs. [...] A complete and detailed project for the organisation of life and space, with the largest possible role for self-management [...] This kind of global project, which is the route rather than the programme, plan or model, bears on collective life and can only be a collective oeuvre which is

simultaneously practical and theoretical. It can depend neither on a party nor on a political bloc; it can only be linked to a diversified, qualitative ensemble of movements, demands and actions.

(Lefebvre, 1976, p. 119)

Who decides and who provides?

As we have explored throughout this chapter, the similarity of autonomous progressive development to a dialectical materialist process of self-management suggests provocative possibilities for the development of economically sustainable social relations. Here the trajectory of these comparisons return full circle to our original interrogation of Lefebvre's spatial appropriation of Marx's dialectical materialism and Turner's critical questioning of *who decides?* (1976, pp. 11–13). This distinction is at the heart of Turner's advocacy for use value as opposed to exchange value, and the necessity for individuals to decide on their own needs and priorities if we are to pursue truly sustainable and progressive development:

> Those who recognise the fact that use-values lie in the relationships between people and things – and not in things themselves – will recognise the significance of alternative means by which alternative ends are sought. This is the issue of economy. If primary values and ends are functional and defined by performance (that is, use rather than quantities), then economy must have as much to do with the means of production, as with productivity. [...] Those who confuse economy with material productivity make a dangerous error. Like market-values, industrial production has its uses but these must be limited or industrialisation will destroy mankind even more surely than the primitive capitalism that generated it.
>
> (Turner, 1976, p. 154)

Based on the observations and comparisons within this chapter we can begin to interpret informal settlements as a concrete example of the practice of autonomy, freedom, and active social participation that Lefebvre advocated in the notion of autogestion and self-management. The most difficult obstacles to such a proposition remain the shift in cultural perception and representation of identity and alterity, autonomy and choice that is suggested by these observations when projected upon a Westernised context (Dovey and King, 2011, pp. 16–18). Both the advocacy for and validation of informal settlements and development practice methodologies suggest the opportunity for a transcendence of capitalist ideology. Yet this proposition is inherently fraught with the necessary recognition of the extreme inequality and deprivation that is so often the foundation of the Global South (Dovey and King, 2012, p. 1031). Thus, unlike historical mis-readings of Turner's work as merely a simplistic advocacy for sweat-equity and self-help, the comparisons and intersections explored in this chapter help to outline the far more spatial and political critique of autonomy and distributed governance:

Those who see this point are bound to recognise the issue of authority which determines the choice of means and which are used to achieve the ends. When economy is understood as resourcefulness, technology is obviously political as it is a matter of who controls resources and their uses. The central issue raised in this book is that of who decides? Who decides, and who provides what for whom is clearly the political issue of power and authority.

(Turner, 1976, p. 154)

This analysis suggests a provocative intersection of spatial theory and spatial practice outlined succinctly in Turner's question of who decides. Questions of who decides and who provides are intrinsically related to Lefebvre's multi-faceted critique of the social relations that produce space. Whilst disparity exists between Turner's anarchist approaches and an institutional interpretation of Marxism, their shared spatial interrogation of authority and power represents a novel and productive interdisciplinary intersection of discourse: not only are their spatial critiques comparable, their points of intersection pose a positive counter-proposition to conventional Western assumptions of space.

Thus when considered in the context of our underlying premise of re-imagining space and architecture, the principles of dialectical materialism provide a spatial foundation and theoretical scale within which to transcend the distinctions of anarchist autonomy and the autogestion of a socio-spatial Marxism. This allows Lefebvre's theoretical advocacy for the social production of space, self-management, and autogestion to be contextualised in explicitly peripheral spaces of informal settlements and economies of absence. Similarly, it allows Turner's contested advocacies for self-help to be re-read and re-imagined outside a purely sweat-equity analysis and positively connected to Lefebvre's explicitly political and theoretical spatial critiques. As such this comparison offers the opportunity to focus upon the articulation of positive difference, alterity, and heteronomy as an intersection between the practical and theoretical discourses of Turner and Lefebvre. It provides an opportunity to re-read Turner's principles of user-choice progressive housing and participatory community practice to confront and contest the qualities and aspirations of Western space.

Turner's concept of housing as a verb seems so disconnected from the reality of contemporary Westernised space, housing, and development as to simply make our comparisons and their implications seem pointless. Yet the positive potentials of an alternative social production of space founded upon principles of user choice, informal space, and political self-management cannot be ignored. It is hard to see any similar 'Lefebvrian' engagement with space or architecture emerging, surviving, or taking hold in the midst of increasingly neoliberal and capitalist Western values.

The condition of Westernised space and social values is no longer a geographical definition, but is now a global condition. Yet in the work of Turner we see a kind of humility and simplicity afforded to the agency of facilitating change that has been shown to succeed in the most challenging circumstances and economies of absence. Perhaps the opportunity remains to find and engage in the contemporary equivalent of these peripheral and alternative extremes. Or perhaps the opportunity exists to manipulate and change the social and political landscape of Western space in such a way as to provide the chance for social and spatial alternatives to emerge. In the next chapter we will look at both the challenges and opportunities of such an aspiration, and pursue an analysis of urbanisation as a concept – from Lefebvre to Massey, and from local social difference to global multiplicity.

Notes

1 This observation is based upon personal correspondence with John Turner, who intriguingly recollects having lived on the same street as Lefebvre in 1970s Paris, yet also noted that he had no knowledge of Lefebvre's discourse or its potential connection to his own work.
2 Hegel's work on dialectic logic itself must be understood in the context of the discourse of Immanuel Kant and Johann Gottlieb Fichte. Writing at the turn of the nineteenth century, both Kant's and Fichte's respective discourses focus on the empirical and rational nature of consciousness and logic (Kant, 2007, p. 26; Lefebvre, 2009, p. 11; Shields, 1999, p. 116).
3 In essence abstracted and philosophical ideas.
4 As explored in the Socratic works of Plato (Kant, 2007, p. 301).
5 Where Kant's and Fichte's processes of logic are bounded and fixed to an internal subject's consciousness, Hegel's interpretation identified contradiction and opposition as being preserved, unified, and elevated within a progressive evolutionary process (Hegel, 2010, p. 33).
6 Namely the first volume of Kapital that coincided with the writing of his thesis on Feuerbach, which Engels was to publish later posthumously (Engels and Marx, 1998; Marx, 2013).
7 Because each contradiction emerges from abstract philosophical thought it necessitates its negative or negation as emerging from an internal conflict and thus the subsequent process of mediation was required to cleanse it and then only to remain a renewed abstract idea.
8 This pursuit of the content and context of relationships ultimately formed the observational framework of historical materialism, leading to Marx's empirical core theories of surplus value, surplus production, and alienation as ways of interpreting the social and political implications of the prevailing capitalist mode of production (Fischer, 1973, pp. 26–28).
9 Engels and Marx, 1998, pp. 106, 154–157.
10 Turner's arrival in Peru coincided with a number of interconnected factors; most notably the political context of 1960s Latin America in general, and specifically Peru's popular socialist democratic government (including communist party support). Fernando Belaúnde Terry (an architect by training) was President of Peru for two non-consecutive terms (1963–1968 and 1980–1985). Deposed by a military coup in 1968, he was re-elected in 1980 after eleven years of military rule. He has been widely recognised for his personal integrity and his commitment to the democratic process.
11 In a rare and notable reference Turner cites E.F. Schumacher's appropriation of Marx's observation that 'the more useful machines there are, the more useless people there will be'.
12 Here Turner is notable for not contesting housing from merely political orientations but as a confrontation of the material, social, and economic inefficiencies that

he saw as impossible to sustain against a finite material world (1976, pp. 46–47). As we shall see throughout our study, the points of comparison drawn between theory and practice, and Global North and South, recur and reveal opportunities to question the values and practices of Westernised space.

13 This observation is similar to the same institutional changes wrought against economically impoverished urban housing in the Global North. Yet here the rampant economic progress of the leading world economies largely masked this re-development under a social imperative. The implications of modernist housing blocks in the UK has been felt by those they were meant to help but who became caught up in the modernist institutionalisation of housing as an object or noun (Hatherley, 2009).

14 In explicit recognition of such observations Turner pointedly cites Edward Sapir, noting how such institutionalisation of housing (and other social productivity) deprives the vast majority of us of the opportunity to engage in the immediate satisfaction of value (Sapir, 1992, p. 321).

15 The project is interlaced with the teaching and research that Professor Alfredo Brillembourg and Professor Hubert Klumpner of UTT undertook in collaboration with ETH Zurich.

16 Critical research into the foundations and interconnected relations of contemporary Western projects is an issue that I reserve the right to explore in detail in future research, but suffice to say, for now it is important to (attempt to) maintain a sense of critical objectivity when citing projects like these.

17 And increasingly in the local periphery (Bauman, 2000; Campbell, 2013).

18 The conventional model leading to capitalistic interpretations and manipulations of housing, centrality and gentrification (Dovey, 2012; Dovey and Sandercock, 2002).

19 Here there are direct comparisons and intersections to be drawn to wider discussion of Doreen Massey's articulations of space and development as not being an inevitability but instead being a rich multiplicity of intersection stories and trajectories explored in later chapters.

20 MOM (Morar de Outras Maneiras / Living in Other Ways): www.mom.arq.ufmg.br

21 This is an increasingly prevalent condition of Western space. It is hoped that research and comparisons like this book might contribute to wider academic and social engagement with the inequality that surrounds us, and yet appears to remain largely hidden in and by the social relations of space.

22 Here we can see early complementary links to later comparisons to the concepts of difference, multiplicity, and the subaltern.

23 Spatial relations of periphery, difference, and alterity are thus here interpreted as the outcome of transgressions generated by the socio-economic necessities of rapid urbanisation and economic migration.

24 In contrast to the inequalities of industrial Manchester the global inequalities of the capitalist mode of production are expressed in the disjunctions between the manufacturing conditions of urban Asia and the consumption of Western states (Harvey, 2010, pp. 162–166; Schumacher, 2011).

25 A city whose origins somewhat mirror that of Milton Keynes in the UK despite the entirely different eventual outcomes.

26 This articulation and political use of 'development' is discussed extensively in chapter four.

27 The pronounced Marxist denunciation of self-help housing raised by Burgess et al. reflects the same institutional Marxist aporia that Lefebvre sought to contest (Lefebvre, 2009, pp. 100–106).

28 Thus any absolute form of politics and ideology cannot be used for the purpose of radical change and re-defined socialism.

References

Abrams, C., 1964. *Housing in the Modern World: Man's Struggle for Shelter in an Urbanising World*. MIT Press, Cambridge, MA.

Abrams, C., 1966. *Squatter Settlements, the Problem and the Opportunity*. Office of International Affairs, Department of Housing and Urban Development, Washington, DC.

Baird, G., 2003. *The Space of Appearance*. MIT Press, Cambridge, MA.

Baltazar, A.P., Kapp, S., 2007. Learning from 'Favelas': The Poetics of Users' Autonomous Production of Space and the Non-ethics of Architectural Interventions. In Proceedings of the International Conference Reconciling Poetics and Ethics in Architecture, McGill University, Canada, September. Available at http://www.arch.mcgill.ca/theory/conference/papers.htm (accessed 1 March 2016).

Baltazar, A.P., Kapp, S., Morado, D., 2008. Architecture as Critical Exercise: Little Pointers Towards Alternative Practices. *Field* 2, 7–30.

Bauman, Z., 2000. *Globalization: The Human Consequences*. Columbia University Press, New York.

Bishwapriya, S., 1994. *Cooperative Autonomy: The Dialectic of State–NGO's Relationship in Developing Countries*. International Institute for Labor Studies, Geneva.

Bower, R., 2016. 'Forgotten Plotlanders: Learning from the Survival of Lost Informal Housing in the UK', *Housing, Theory and Society*.

Brenner, N., 1997. Global, Fragmented, Hierarchical: Henri Lefebvre's Geographies of Globalisation. *Public Culture* 10, 135–167.

Brenner, N., Elden, S., 2009. Introduction, in: *Henri Lefebvre – State, Space, World; Selected Essays*. University of Minnesota Press, Minneapolis.

Brillembourg, A., Klumpner, H. (Eds), 2012. *Torre David: Anarcho Vertical Communities*. Lars Müller, Zurich.

Bromley, R., 2003. Peru 1957–1977: How Time and Place Influenced John Turner's Ideas on Housing Policy. *Habitat International* 27, 271–292.

Burgess, R., 1978a. Self-Help Housing. A New Imperialist Strategy? A Critique of the Turner School. *Antipode* 9, 50–60.

Burgess, R., 1978b. Petty Commodity Housing or Dweller Control. *World Development* 6, 1105–1133.

Burgess, R., 1982. Self-help Housing Advocacy: A Curious Form of Radicalism. A Critique of the Work of John F.C. Turner, in: Ward, P.M. (Ed.), *Self-Help Housing: A Critique*. Mansell, London.

Burgess, R., 2001. The Compact City Debate: A Global Perspective, in: Burgess, R., Jenks, M. (Eds), *Compact Cities: Sustainable Urban Forms for Developing Countries*. Routledge, London.

Burgess, R., Carmona, M., Kolstee, T. (Eds), 1997a. *The Challenge of Sustainable Cities: Neoliberalism and Urban Strategies in Developing Countries*. Zed Books, London.

Burgess, R., Carmona, M., Kolstee, T., 1997b. Contemporary Policies for Enablement and Participation: A Critical Review, in: Burgess, R., Carmona, M., Kolstee, T. (Eds), *The Challenge of Sustainable Cities: Neoliberalism and Urban Strategies in Developing Countries*. Zed Books, London.

Campbell, P., 2013. Collateral Damage? Transforming Subprime Slum Dwellers into Homeowners. *Housing Studies* 28, 453–472. doi:10.1080/02673037.2013.759543

Chang, H.-J., Grabel, I., 2014. *Reclaiming Development: An Alternative Economic Policy Manual*, 2nd Edition. Zed Books, London.

Coetzer, N., 2010. Towards a Dialogical Design Studio: Mediating Absurdities in Undergraduate Architectural Education in South Africa. *South African Journal of Art History* 25, 101–117.

Cruz, T., 2012. Mapping Non-Conformity: Post-Bubble Urban Strategies. *Hemispheric Institute E-Misférica*. Available at http://hemi.nyu.edu/hemi/en/e-misferica-71/ cruzhttp://hemisphericinstitute.org/hemi/en/e-misferica-71/cruz (accessed 1 March 2016).

de Carlo, G., 1949. The Housing Problem in Italy. *Freedom*, 12 June, 19 June.

Dovey, K., 2012. *The Temporary City.* Routledge, London.

Dovey, K., 2013. Informalising Architecture: The Challenge of Informal Settlements. *Architectural Design* 83, 82–89.

Dovey, K., King, R., 2011. Forms of Informality: Morphology and Visibility of Informal Settlements. *Built Environment* 47, 11–29. doi:10.2148/benv.37.1.11

Dovey, K., King, R., 2012. Informal Urbanism and the Taste for Slums. *Tourism Geographies* 14, 275–293. doi:10.1080/14616688.2011.613944

Dovey, K., King, R., 2013. Interstitial Metamorphoses: Informal Urbanism and the Tourist Gaze. *Environment and Planning D: Society and Space* 31, 1022–1040.

Dovey, K., Sandercock, L., 2002. Hype and Hope. City: Analysis of Urban Trends, Culture, Theory, *Policy, Action* 6, 83–101.

Elden, S., 2004. *Understanding Henri Lefebvre: Theory and the Possible.* Continuum, London.

Elden, S., 2008. Mondialisation Before Globalisation, in: Goodewardena, K., Kipfer, S., Milgrom, R., Schmid, C. (Eds), *Space, Difference, Everyday Life: Reading Henri Lefebvre.* Routledge, New York.

Engels, F., 1844. *The Condition of the Working Class in England.* Otto Wigand, Leipzig.

Engels, F., Marx, K., 1975. The Condition of the Working Class in England, in: *The Collected Works of Marx and Engels.* International Publishers, New York.

Engels, F., Marx, K., 1998. *The German Ideology: Including Theses on Feuerbach and an Introduction to the Critique of Political Economy.* Prometheus, New York.

Esteva, G., 2010. Development, in: Sachs, W., *The Development Dictionary.* Zed Books, London.

Fernández-Maldonado, A.M., 2007. Fifty Years of Barriadas in Lima: Revisiting Turner and De Soto. Paper presented at the ENHR 2007 International Conference on Sustainable Urban Areas. Rotterdam, the Netherlands.

Fichte, J.G., 2000. *Foundations of Natural Right. Cambridge Texts in the History of Philosophy.* Cambridge University Press, Cambridge.

Fichter, R., Turner, J.F. (Eds), 1972. *Freedom to Build.* Macmillan, New York.

Firth, R., 2012. *Utopian Politics: Citizenship and Practice.* Routledge Innovations in Political Theory. Routledge, Abingdon.

Fischer, E., 1973. *Marx in His Own Words.* Pelican Books, London.

Franks, M.A., Turner, J.F.C., 1995. Different Ways of Seeing. *New Economics* (Autumn, Winter, Autumn, Winter), first two in an eight-part series.

Fraser, M., 2012a. The Global Architectural Influences on London. *Architectural Design* 82, 14–21.

Fraser, M., 2012b. The Future is Unwritten: Global Culture, Identity and Economy. *Architectural Design* 82, 60–65.

Friere, P., 1996. *Pedagogy of the Oppressed*, 2nd Edition. Penguin, London.

Geddes, P., 1949. *Cities in Evolution*, 2nd Edition. Williams and Norgate, London.

Goonewardena, K., Kipfer, S., Milgrom, R., Schmid, C., 2008. Globalizing Lefebvre?, in: Goodewardena, K., Kipfer, S., Milgrom, R., Schmid, C. (Eds), *Space, Difference, Everyday Life: Reading Henri Lefebvre*. Routledge, New York.

Gordon, A., 1974. Loose Fit, Low Energy, Long Life. *RIBA Journal*, January, 9–12.

Habraken, J., 1972. *Supports: Alternative to Mass Housing*. Architectural Press, London.

Hall, P., 2013. *Good Cities, Better Lives: How Europe Discovered the Lost Art of Urbanism*. Routledge, London.

Harms, H., 1976. Limitations of Self-Help. *Architectural Design* 46, 230–231.

Harms, H., 1982. Historical Perspectives on the Practice and Politics of Self-Help Housing, in: Ward, P.M. (Ed.), *Self-Help Housing: A Critique*. Mansell, London.

Harris, R., 1998. The Silence of the Experts: 'Aided Self-Help Housing' 1939–1954. *Habitat International* 22, 165–189.

Harris, R., 1999. Slipping Through the Cracks: The Origin of Aided Self-Help Housing 1918–1953. *Housing Studies* 14, 281–309.

Harris, R., 2003. A Double Irony: The Originality and Influence of John F.C. Turner. *Habitat International* 27, 245–269.

Harvey, D., 2003. The Right to the City. *International Journal of Urban and Regional Research* 27, 939–941.

Harvey, D., 2010. *The Enigma of Capital and the Crises of Capitalism*. Oxford University Press, Oxford.

Harvey, D., 2012. *Rebel Cities*. Verso, London.

Hatherley, O., 2009. *Militant Modernism*. Zero Books, New York.

Hegel, G.F., 2010. *George Friedrich Hegel and the Science of Logic*. Cambridge University Press, Cambridge.

Henrich, D., Pacini, D.S., Green, G.W., 2005. Between Kant and Hegel. *The Review of Metaphysics* 59, 423–425.

Hill, D., 2015. *Dark Matter and Trojan Horses: A Strategic Design Vocabulary*. Strelka Press, Moscow.

Hodkinson, S., 2012. The Return of the Housing Question. *Ephemera: Theory and Politics in Organization* 12, 423–444.

Hughes, J., Sadler, S., 1999. *Non-Plan: Essays on Freedom, Participation and Change in Modern Architecture and Urbanism*. Routledge, London.

Illich, I., 1976. *Deschooling Society*. The Philips Park Press, Manchester, UK.

Ingham, B., 1993. The Meaning of Development: Conversations Between 'New' and 'Old' Ideas. *World Development* 21, 1803–1821.

Jenkins, P., Milner, J., Sharpe, T., 2009. A Brief Historical Review of Community Technical Aid and Community Architecture, in: Jenkins, P., Forsyth, L. (Eds), *Architecture, Participation and Society*. Routledge, London.

Jiron, P., 2010. The Evolution of Informal Settlements in Chile: Improving Housing Conditions in Cities, in: Hernandez, F., Kellett, P., Allen, L.K. (Eds), *Rethinking the Informal City*. Berghahn Books. New York.

Kant, I., 2007. *Critique of Pure Reason*, New Revised Edition. Penguin, London.

Kipfer, S., Goonewardena, K., Schmid, C., Milgrom, R., 2008. *On the Production of Henri Lefebvre*. Routledge, New York.

Kotanyi, A., Vaneigem, R., 1961. Elementary Program of the Bureau of Unitary Urbanism, in: Knabb, K. (Ed.), *Situationist International Anthology*. Bureau of Public Secrets, Berkeley, CA.

Kristin, R., 1988. *The Emergence of Social Space: Rimbaud and the Paris Commune*. Macmillan, Basingstoke.

Kusno, A., 2000. *Behind the Postcolonial: Architecture, Urban Space, and Political Cultures in Indonesia.* Architext Series, Routledge, London.

Lefebvre, H., 1968a. *Dialectical Materialism.* Jonathon Cape, London.

Lefebvre, H., 1968b. *The Sociology of Marx.* Allen Lane, London.

Lefebvre, H., 1969. *The Explosion.* Monthly Review Press, New York.

Lefebvre, H., 1976. *The Survival of Capitalism.* Allison and Busby, London.

Lefebvre, H., 1980. Marxism Exploded. *Review* 4, 19–32.

Lefebvre, H., 1991. *The Production of Space.* Blackwell Publishing, Oxford.

Lefebvre, H., 1996. *Writings on Cities.* Blackwell Publishing, Oxford.

Lefebvre, H., 2003. *The Urban Revolution.* University of Minnesota Press, Minneapolis.

Lefebvre, H., 2009. *Henri Lefebvre – State, Space, World: Selected Essays.* University of Minnesota Press, Minneapolis.

Mangin, W., 1967. Latin American Squatter Settlements: A Problem and a Solution. *Latin American Research Review* 2, 65–98.

Marx, K., 2013. *Capital: Volumes One and Two.* Wordsworth Classics of World Literature. Wordsworth, London.

Merrifield, A., 2006. *Henri Lefebvre – A Critical Introduction.* Routledge, New York.

Merrifield, A., 2013a. Citizen's Agora. *Radical Philosophy* 179, 31–35.

Merrifield, A., 2013b. *The Politics of the Encounter: Urban Theory and Protest Under Planetary Urbanization (Geographies of Justice and Social Transformation).* University of Georgia Press, Athens, GA.

Nalbantoglu, G.B., Wong, C.T. (Eds), 1997. *Postcolonial Space(s).* Princeton Architectural Press, New York.

Neuwirth, R., 2006. *Shadow Cities: A Billion Squatters, A New Urban World,* New Edition. Routledge, New York.

Nicholson, S., 1972. The Theory of Loose Parts, An Important Principle for Design Methodology. *Home* 4, 5–14.

Peattie, L., Doebele, W., 1973. Review of *Freedom to Build. Journal of the American Institute of Planning* 39, 66–67.

Pugh, C., 2000. Squatter Settlements: Their Sustainability, Architectural Contributions, and Socio-economic Roles. *Cities* 17, 325–337.

Puttnam, R.D., Leonardi, R., Nanetti, R.Y., 1993. *Making Democracy Work: Civic Traditions in Modern Italy.* Princeton University Press, Princeton, NJ.

Rapoport, A., 1987. Spontaneous Settlements as Vernacular Design, in: Patton, C. (Ed.), *Spontaneous Shelter: International Perspectives and Prospects.* Temple University Press, Philadelphia, PA.

Roy, A., 2005. Urban Informality: Toward an Epistemology of Planning. *Journal of the American Planning Association* 71, 147–158.

Roy, A., 2011. Slumdog Cities: Rethinking Subaltern Utopianism. *International Journal of Urban Regional Research* 35, 223–238.

Sachs, W., 2010. One World, in: Sachs, W., *The Development Dictionary.* Zed Books, London.

Sandercock, L., 1998. *Making the Invisible Visible: A Multicultural Planning History.* University of California Press, Berkeley.

Sanyal, B., Vale, L.J., Rosan, C., 2008. Four Conversations, in: Sanyal, B., Vale, L.J., Rosan, C. (Eds), *Planning Ideas That Matter: Livability, Territoriality, Governance and Reflective Practice.* MIT Press, Cambridge, MA.

Sapir, E., 1992. *Selected Writings in Language, Culture, and Personality,* New Edition. University of California Press, Berkeley.

Schon, D., 1987. Institutional Learning in Shelter and Settlement Policies, in: Rodwin, L. (Ed.), *Shelter, Settlement and Development*. Allen and Unwin, Boston, MA.

Schumacher, E.., 2011. *Small is Beautiful – A Study of Economics as if People Mattered*, New Edition. Vintage, London.

Shields, R., 1999. *Lefebvre, Love & Struggle: Spatial Dialectics*. Routledge, London.

Stepelevich, L.S., 2008. Philosophie als System bei Fichte, Schelling und Hegel. *Journal of the History of Philosophy* 15, 485–487.

Turner, J.F.C., 1967. Barriers and Channels for Housing Development in Modernizing Countries. *Journal of the American Institute of Planning* 33, 167–181.

Turner, J.F.C., 1968. The Squatter Settlement: Architecture that Works. *Architectural Design* 38, 355–360.

Turner, J.F.C., 1972a. Housing as a Verb, in: Fichter, R., Turner, J.F. (Eds), *Freedom To Build*. Macmillan, New York.

Turner, J.F.C., 1972b. The Re-Education of a Professional, in: Fichter, R., Turner, J.F. (Eds), *Freedom To Build*. Macmillan, New York.

Turner, J.F., 1974. The Fits and Misfits of People's Housing. *RIBA Journal* 81(2), 12–21.

Turner, J.F., 1976. *Housing by People: Towards Autonomy in Building Environments*. Marion Boyars, London.

Turner, J.F., 1979. Housing in Three Dimensions: Terms of Reference for the Housing Question Redefined, in: Bromley, R. (Ed.), *The Urban Informal Sector: Critical Perspectives on Employment and Housing Policies*. Pergamon Press, Oxford.

Turner, J.F., 1983. From Central Provider to Local Enablement. *Habitat International* 7, 207–210.

Turner, J.F., 1986. Future Directions in Housing Policy. *Habitat International* 10, 7–25.

Turner, J.F., 1992. Foreword, in: Mathey, K. (Ed.), *Beyond Self-Help Housing*. Mansell, London.

Turner, J.F., 1996. Tools for Building Community. *Habitat International* 20, 339–347.

Turner, J.F., 1997. Learning in a Time of Paradigm Change, in: Burgess, R., Carmona, M., Kolstee, T. (Eds), *The Challenge of Sustainable Cities*. Zed Books, London.

Vega-Christie, D., Crane, J., 1948. Peru Establishes a National Housing Corporation. *American City* 43, 96–97.

Ward, C., 1972. Preface, in: Fichter, R., Turner, J.F. (Eds), *Freedom To Build*. Macmillan, New York.

Ward, C., Hardy, D., 1972. Plotlanders. *Oral History Journal* 13, 57–70.

Ward, P.M. (Ed.), 1982. *Self-Help Housing: A Critique*. Mansell, London.

Ward, P.M. (Ed.), 1999. *Colonias and Public Housing Policy in Texas and Mexico: Urbanisation by Stealth*. University of Texas Press, Austin.

Ward, P.M., 2004. Informality of Housing Production at the Urban–rural Interface: The Not-so-Strange Case of Colonias in the US – Texas, the Border and Beyond, in: Roy, A., AlSayyad, N. (Eds), *Urban Informality*. University of California Press, Berkeley.

Ward, P.M., 2005. The Lack of 'Cursive Thinking' with Social Theory and Public Policy: Four Decades of Marginality and Rationality in the so-called 'Slum', in: Roberts, B., Wood, C. (Eds), *Rethinking Development in Latin America*. Pennsylvania State University Press, Philadelphia.

Ward, P.M., 2008. Self-Help Housing Ideas and Practice in the Americas, in: Sanyal, B., Vale, L.J., Rosan, C. (Eds), *Planning Ideas that Matter: Livability, Territoriality, Governance and Reflective Practice*. MIT Press, Cambridge, MA.

World Bank, 1997. *World Development Report 1997: The State in a Changing World*. World Bank, Washington, DC.

2 Spatial relations, difference, and multiplicity

In the last chapter we applied a critical lens to the respective discourses of John Turner and Henri Lefebvre. This analysis highlighted connections and comparisons between these two protagonists such as autonomous user-defined housing and dialectical materialism, sites and services housing models and autogestion, and grass-roots development with the social relations of production. These points of comparison afforded an opportunity to re-read and re-imagine Lefebvrian spatial advocacies as having been realised by Turner's apparently anarchist principles of practical spatial agency. This analysis subsequently posed a critical counter-narrative to the prevailing neoliberal and capitalist social relations of Westernised space: is anarchistic grass-roots development the most practical spatial methodology for achieving the kinds of space advocated by Marxist and socialist thinkers like Lefebvre?

Building upon chapter one, in this second chapter we set out to explore how Lefebvre's critical socio-spatial theory can be connected to a more explicitly contemporary and globalised context. This analysis specifically seeks to draw a new theoretical lineage from Lefebvre's spatial theory to the discourse of Doreen Massey.

This connection is vital to the wider premise of our book as it builds upon our initial understanding of space as both a material and dialectic process. With the connection to Massey we explicitly expand the implications of space as a social process to questions of global urbanisation, appropriation, difference, and positive multiplicity. This expansion of our theoretical framework seeks to connect Lefebvre's post-Marxist socio-spatial analysis to the post-structural spatial theory found in Massey's discourse. In essence, in this chapter we seek to re-imagine concepts of space, identity, and difference as explicitly inter-connected and interdependent with our critical comparison of the abstract formality of Westernised space and the informal development of the Global South.

This trajectory of research re-contextualises the spatial critiques of Massey and Lefebvre by specifically connecting their respective advocacies for the positive potential of space. This comparison observes intersections in their discourses concerning the social and political relations that our conceptions and experience of space are ridden with – Lefebvre in 'space as a social product' (1991, p. 26) and reflectively in Massey's 'relationality of space' (2005, pp. 100, 194). Thus, our premise is to explore aspects of this critical intersection within the key texts of both theorists, as well as connections and implications to the wider post-structural field of spatial theory.

From the outset both Lefebvre's and Massey's discourses can be perceived as built upon political foundations drawn from Marxist and socialist conceptions of space and the fundamentals of dialectical reasoning and process (Lefebvre, 1980, p. 23; Massey, 2007, pp. 66–67). However our comparison suggests that Massey's discourse offers a global contextualisation of Lefebvre's observations of differential spaces, and a post-structural critique of the relational construction of space as expressed in the concept of positive multiplicity (Saldanha, 2013, pp. 48–49). In observing this lineage as projecting from Lefebvre's discourse on 'the right to the city' (Harvey, 2003; Lefebvre, 1996, p. 148) and 'the right to difference' (Lefebvre, 2003, p. 96), we will begin to frame points of intersection with Massey's advocacy of participation, appropriation, and positive relational multiplicity (2004, pp. 14–15). This connection of Marx's, Lefebvre's, and Massey's interpretations of socio-spatial relations also implicitly resonates with urban questions and problems confronted in the participatory development practices of Turner and Hamdi.

The trajectory of our comparisons also reflects the academic and socio-political transition from structuralism to post-structuralism that defined the historical context from which both Lefebvre's and Massey's discourses emerged. As we will discover, Lefebvre's advocacy of differential space offers a clear connection to post-structural theory. Yet contemporary academic utilisations of such differential space remains largely abstract in advocacies of embodiment, spontaneity, and moments as emergent events, rather than offering positive practical examples or methodologies of how to achieve such space (Goonewardena et al., 2008a, p. 292; Shields, 1999, p. 183). By contextualising Lefebvre's spatial critique against Massey's post-structural discourse this chapter confronts his articulations of differential space and appropriation with the questions of global inequality that pervade post-colonial theory and development discourse. Thus our explicit pursuit of the connection between Lefebvre's concept of differential space and Massey's advocacy of the multiplicity of space provides a new *interpretation of*, not *addition to*, Lefebvre's differential space, specifically by exploring global and post-colonial contexts of alternative, informal, and participatory development.

We will observe that whilst Lefebvre advocates the interdependence of time and space (2009, p. 40), Massey's critique of the social and relational production of space is far more extensively grounded upon a critique of global geographies of inequality and development (2009, p. 22). Thus in Massey we find a theoretical projection of Lefebvre and Marxist theory that allows a critical interrogation of space, architecture, and development in a contemporary global context. This offers a conception of Lefebvre's critique of space that is open-ended, relational, plural, and (most importantly) positive. It is not a negative critique of space and difference, but an advocacy of valuing and working within the positive relational specificity and happenstance – the 'throwntogetherness' of space (2005, pp. 140–141).

Massey's discourse utilises a series of propositions for space, each built around the imperative of multiplicity: the recognition of other and alternative interpretations of the world as part of the relations that exist within space (and time). This recognition of the multiplicity of space reinforces our research methodology of drawing development practices into critical comparison with further theoretical trajectories of post-colonialism and subalterneity in later chapters. Thus, the lineage from structuralism to post-structuralism and Lefebvre to Massey acts as a foundation for later trajectories into elements of the discourses of Derrida (2001, 1996, 1987), Bhabha (2006, 2004), and Spivak (1999, 1998, 1985, 1984), each built upon the implications and values of space and difference.

This is not a unique proposition. Yet the trajectory via Lefebvre and Massey is new, and provides novel opportunities for comparison and re-imagination of Westernised space and architecture in the context of development theory. This new critical connection becomes valuable because of its articulation of space and spatial relations as a medium of multiplicity and participation (Massey, 2005, p. 12). This critical grounding and framework supports our wider research premise and questions the political and economic potential of space as framed and defined by social relations and contextualised material practices of cities, difference, and multiplicity:

> By seeking to point the way towards a different space, towards the space of a different (social) life and of a different mode of production, this project straddles the breach between science and utopia, reality and ideality, conceived and lived. It aspires to surmount these oppositions by exploring the dialectical relationship between 'possible' and 'impossible', and this both objectively and subjectively.
>
> (Lefebvre, 1991, p. 60)

This re-imagining of difference as a positive and imperative aspect of space is crucial to the wider premise of this book. The theoretical foundations outlined in this chapter are necessary in order to validate and ground the comparisons of space and architecture drawn from the Global North and South. Without re-imagining the relational interdependence of space, difference, and multiplicity how can exemplars of development practice drawn from contexts of extreme economic absence be critically (and practicably) compared to the neoliberal affluence of the Global North? If we are to re-imagine space and architecture, and if we are to learn from other ways of engaging with people found in development practice, then we have to begin by recognising that the right to the city is too simple an answer. The right to difference – spatial transgressions, appropriations, and agency – and the agency of difference are integral to the premise of space as a relational multiplicity – as something that is open to the kinds of positive real-world agency and change observed in pro-poor and grass-roots development practice.

The Production of Space

Since Donald Nicholson-Smith's English translation of *The Production of Space* (1991) there have been multiple insightful analyses of Lefebvre's perhaps most studied text. These include notable works by Andy Merrifield, Stuart Elden, Rob Shields, Neil Brenner, Lukasz Stanek, and Kanisha Goonewardena et al., many of which are utilised in this book. However, it is explicitly not the intention of our book to untangle Lefebvre's discourse in its entirety, or even specific arguments. Our aim is to extract several key spatial and critical concepts and methodologies from this discourse and contextualise them within our attempts to re-imagine space and architecture via the post-structural comparison of the Global North and South.

The evolution of Lefebvre's discourse is remarkably complex, broad, and nuanced. As Merrifield notes, '*The Production of Space* was Lefebvre's fifty-seventh book' (2006, p. 99), and marks a point of consolidation and realisation of much of his earlier propositions written in the wake of the 1968 Paris political protests and his abrupt break from the French Communist Party and the Situationists (Sadler, 1999, p. 45). This influential text explores the spatial implications of concepts of representation, dialectics, spontaneity, everyday life, political struggle, and philosophy, amongst many others. Similarly, the ranges of theoretical discourse upon which Lefebvre draws are further suggestive of its complexity, including Hegel, Marx, Kostas Axelos, Michel Foucault, and Friedrich Nietzsche (Elden, 2004, p. 73). It also marks somewhat of a coalescence of many strands of Lefebvre's discourse into a single work, seeking to reveal the connections between the urban process, spatial relations, politics, and economics. This coalescence is summarised within the now famous conception that (social) space is a (social) product (Lefebvre, 2009, pp. 186–187; Schmid, 2008, p. 28).

The articulation of space as contingent upon the social relations that define its 'production' is Lefebvre's reaction to the fragmentation of space that he observes in the academic, political, and bureaucratic abstractions of space.[1] *The Production of Space* offers an explicit critique of this abstraction, analysing it with the now seminal spatial framework and critical lens of a trialectic relational model: physical space (nature), mental space (abstractions of space), and social space (the space of human interaction) (Lefebvre, 1991, pp. 38–39; Merrifield, 2006, p. 103). In essence these three interdependent layers of socio-spatial relationships all combine to produce Westernised space. Whilst this exact same analysis of relationships can be applied to informal space and the Global South, these relationships are also articulated and experienced in very different ways.

Utilising this critical lens, Lefebvre is explicit about the implications of the connection of the production and manipulation of space, describing the progression from absolute to historical and industrial space, on to contemporary abstract spaces and homogenous global urbanisation (Lefebvre, 1991, pp. 46–49; Merrifield, 2006, p. 130). For Lefebvre, space is abstracted and fragmented because of its dichotomy with the representation of the political and economic relations that underlie capitalist space as logical cohesion (Shields, 1999, p. 146). Merrifield succinctly paraphrases the implications of the fragmentation and subsequently induced hegemonic abstract space that 'tends to sweep everybody along, molding people and places in its image, incorporating peripheries as it peripherises centres, being at once deft and brutal, forging unity out of fragmentation' (2006, p. 112). In light of this critique Lefebvre's expansion of the analytical methodology of the dialectic into a spatial trialectic (2009, pp. 262–265) provides a paradigmatic confrontation of this fragmented space, and reveals the now widely observed and contested interdependent relations of conceived, perceived, and lived space (1991, pp. 33, 38–39).

In contrast to conventional structural abstractions of space, the continuous and open-ended instability of such a trialectic spatial framework does not provide a static resolution to space, as to do so would invalidate the political potential of social production and return space to mere ideology (Elden, 2004, p. 23). Instead the material dialectic process is spatially re-appropriated by Lefebvre, creating a fluid, participatory, and open-ended interpretation of space as inseparable from practice – space as a process with three specific moments that intersect, overlap, and blur into each other (Shields, 1999, p. 116).

Examples of the production and trialectic relations of space are sometimes surprisingly difficult to pinpoint and pin down. In architecture the most obvious examples appear to be either historical (religious buildings are a favourite) or negative (shopping centres, military spaces, gated communities, etc.). Avoiding these negative frameworks we might instead cite examples seemingly from outside architecture, in this case allotments.

Allotments are a fascinating example of the social production of space (Crouch and Ward, 1997). They represent an uncommonly (socially, economically, politically) positive and proactive engagement with space.[2] They are produced by social relations that stand counter to pervading capitalist space: the exchange of your own time to produce food rather than paying for twenty-first century convenience. The everyday lived reality engages with a different appreciation of time and seasonality.[3] It is a bodily connection with space and earth and the slow rhythms of food and (informal/anarchist) community. Allotments are perceived as a counter narrative and conceived of as a marginal and conventionally out-dated pastime. Yet in spite of recent attempts at capitalist commodification and reification they remain (for now) a fixed aspect of UK (and wider Western) culture that reflects different (and positive) social relations and the production of alternative spaces.

Lefebvre utilises his trialectic observations of the relational production of space to reveal the complex spatial manifestations of economic and political influence upon the space of everyday life (1991, p. 34). It is this positive duality of space as relationally constructed in an interlaced trialectic and yet also inherently unfinished and continuous that reinforces Lefebvre's advocacy of social space. The action, moments, and spontaneity of space and practice allow people to make 'political purchase of process thinking, of conceiving reality in fluid movement, in its momentary existence and transient nature' (Merrifield, 2006, p. 104).

In spite of the implications of the contemporary capitalist abstraction of space, the possibility to perceive space as forever unfinished remains provocative. Lefebvre's discourse re-imagines space with an implicit advocacy of the positive potential of space and social relations as a continuous trialectic. Yet what remains to be identified are the positive practical implications and methodologies for confronting the complex economic bureaucracies and systems of political power manifest within the superstructure of complex societies and cities (Schmid, 2008, p. 43). Thus, this chapter seeks to explore a re-articulation of Lefebvre's Western spatial critique that connects abstract space with a global and post-structural Marxist critique,[4] and of the social relations of capitalist hegemony.

> Contradictions of capitalism henceforth manifest themselves as contra-dictions of space. To know how and what space internalizes is to learn how to produce something better, is to learn how to produce another city, another space, a space for and of socialism. To change life is to change space; to change space is to change life. Neither can be avoided.
>
> (Merrifield, 2006, p. 108)

This dialectic proclamation to change life and space remains one of the most powerful evocations of the necessity of theory and practice to be entwined in the production of space. Yet Lefebvre's spatial trialectic implies a continuously evolving and dynamic construct of social relations that inform and compose social space. Thus the production of social and spatial relations proceeds without a need to totalise and resolve the complexity of space, instead only aspiring towards a post-structural desire to engage with the interconnected relations that produce it:

> Critical knowledge has to capture in thought the actual process of pro-duction of space. [...] It is a task that necessitates both empirical and theoretical research, and it's likely to be difficult. It will doubtless involve careful excavation and reconstruction; warrant induction and deduction; journey between the concrete and the abstract, between the global and the local, between self and society. Between what's possible and what's impossible.
>
> (Merrifield, 2006, p. 108)

As noted above, the question of how to produce different or alternative spaces is intimately connected to the dialectics of the concrete and abstract, self and society, possible and impossible, formal and informal. Urban space as the richest and densest expression of the spatial and relational trialectic becomes for Lefebvre the medium in which such a contestation of the social and political production of space is most clearly expressed. The city's complexity and density reveals and highlights the abstract nature of passive social space and provokes Lefebvre's ground-breaking rallying cry for 'the right to the city' (1996, p. 148). This contestation and confrontation of the city has been continually renewed throughout many major contemporary spatial discourses – from David Harvey (2003), Saskia Sassen (2001), Iain Borden (2001), Mike Davis (2007), Robert Neuwirth (2006), and so on – and will form the foundation for our overall trajectory towards Massey's discourse on the positive multiplicity of space.

The right to the city

Advocacy of a Marxist political awakening of the working class to their 'right to the city' begins with Engels' observations of worker housing conditions and social conditions in nineteenth century Manchester (1975).[5] Subsequently the modern post-industrial city has long been recognised as the point of aggregation, debate, and conflict that most sharply reflects the political and economic manifestations of class inequality and power:

> For the working class, victim of segregation and expelled from the traditional city, deprived of a present and possible urban life, a practical problem owes itself, a political one, even if it hasn't been posed politically, and even if until now the housing question [...] has masked the problematic of the city and the urban.
>
> (Lefebvre, 1991, p. 100)[6]

This 'problematic of the city and the urban' is a conception that allows Lefebvre to explicitly implicate the urban in class struggle (2003, pp. 7, 37). His critique of the social production of social space is thus implicated with the challenges of the post-industrial city identified in contemporary critiques of capitalist urbanisation (Bauman, 2000, pp. 35–37; Harvey, 2012, p. 29; Massey, 2007, pp. 17, 84). Massey's contribution to this advancement is an engagement with the material inequality of global space and cities, and the interrogation of the relational contradictions that define contemporary abstract urban space (Massey, 2005, p. 103). Yet even in this contemporary global context 'the right to the city' is still inherently an advocacy of the social production of urban space through appropriation and contestation. It advocates engagement with the city as a means to expose the underlying inequalities and contradictions of abstract space and capitalism:

> Marx [...] held that we CHANGE ourselves by changing our world and vice versa. This dialectical relation lies at the root of all human labor. [...] We are, all of us, architects, of a sort. We individually and collectively make the city through our daily actions and our political, intellectual and economic engagements. But in return, the city makes us.
>
> (Harvey, 2003, p. 939)

The idea here advanced by David Harvey is a very succinct summation of the 'the right the city' as a concept that exists in both Marx and Lefebvre. It explicitly reflects both the negative implications of abstract urban space, and the positive spatial advocacy of Lefebvre (2009, pp. 167, 194, 210, 1996, pp. 169–170) and Massey (1992, p. 77). The dialectic that *we make the city* but in return *the city makes us* can be considered as a simplified recitation of a spatial trialectic but more so as an expression of the materialist unfolding of history and the inherent ability to change things for the better.[7] This dialectic and relational identity of urban space explicitly relates to Lefebvre's advocacy of the potential of the city as being realised only through the participation and production of social relations that define the lived reality of cities (1996, p. 155), and in what Lefebvre describes as the 'exquisite oeuvre of praxis and civilisation' (1996, p. 126):

> The right to the city manifests itself as a superior form of rights: right to freedom, to individualisation in socialisation, to habitat and to inhabit. The right to the oeuvre, to participation and appropriation (clearly distinct from the right to property), are implied in the right to the city.
>
> (Lefebvre, 1996, p. 173)

The intentional and specific references here to participation and appropriation advance Lefebvre's advocacy in a very explicit way. The right to the city can only be perceived as a positive force when it is understood as linked to the necessity for active engagement in the social relations that produce our space (Lefebvre, 1996, p. 88; Shields, 1999, p. 35). Thus we find explicit reference to both conflict and contestation in Harvey's recent discourse on *Rebel Cities* (2012, pp. 151–153) and Massey's *World City* (2007, p. 188), each attempting to transcend retrospective critiques of the abstract nature of capitalist cities towards positive alternative practice.

Examples of spatial engagements with 'the right to the city' reflect the complexity of the power and spatial relations at play in what appears to be a simple concept. The various Parisian contestations of the political contexts, notably the 1871 Paris commune or the May 1968 political protests are commonly cited examples. Perhaps most commonly cited in reference to Lefebvrian theories of space, architecture are the actions and agency of the Situationist International movement (Debord, 1994; Knabb, 2006; Sadler, 1999), though the long-term positive impact of these events remains less than clear.

Similarly we could look to the urban protests that defined the contemporary Arab Spring uprisings and civil unrest – Tunisia, Egypt, Libya, Yemen, Bahrain, Syria – noting the repeated use of the citizen occupation of city squares (Egypt's Tahrir Square translates as *Liberation* Square)[8] as a means to provoke political unrest and (apparent) change. We might also highlight the noted lack of any squares in Egypt's new political city centre (Elba, 2015), echoing the manipulation of urban fabric to supress political protest first realised by Haussmann in Paris (Mumford, 1961, pp. 418–460).

Explicitly positive and agonistic (not antagonistic or revolutionary)[9] spatial engagements with the spatial right to the city remain difficult to define. The various 'Occupy Movements' were specifically urban activations of civic rights and political disagreement, yet they were polemical and lacked any agency beyond protest. The right to protest is increasingly difficult to realise as a fundamental principle of the citizens' 'right to the city'. This is perhaps best highlighted in the UK by the changing rules about protesting in Parliament Square, which now require advanced written permission. This subtle shift in political freedom was so starkly (yet equally ephemerally) highlighted in Mark Wallinger's Turner prize winning[10] art installation, 'The State of Britain' at the Tate Britain in 2007.[11]

Conflict is often revealed in space and protest as inherent in the participation and contestation of the right to the city (Lefebvre, 1969, pp. 68–70). Appropriation, transgressions, and conflicts are expressions of *difference* that are implicitly alien to the assumed coherence of capitalist space and cities that rely on the logical cohesion of market forces. As we will see in the next chapter, this question of transgressions and capitalist cohesion is implicit within Lefebvre's discourse and more recent contemporary appropriations of his theory (Lefebvre, 2009, pp. 69–77, 120).

For Lefebvre, the right to challenge, contest, and remake the social relations of space inherently implies the creation of 'differential spaces' which are identified as counter, peripheral, and alternative to formal and conventional space. Such spaces are the expression of different social relations and the different political and economic attitudes that such positive difference and alterity might engender (Harvey, 2003, p. 939). Yet within these distinctions are continued paradoxes. Interdependent with the possibility for social change and social space that is entailed by the right to the city is our own collective responsibility and culpability within such complex and interdependent variables of space, economics, and society, and the implication of the social inequality of difference:

> The right to the city therefore signifies the constitution or reconstitution of a spatio-temporal unit, of a gathering together instead of a fragmentation. It does not abolish confrontations and struggles. On the contrary!
>
> (Lefebvre, 1996, p. 195)

Lefebvre's critique of the homogenous, abstract, and ideological representations of spaces is clearer to perceive in contrast with alternative and differential space. Simply put, differential spaces express and articulate social conflicts, contestations, and possibilities (Shields, 1999, p. 183). They are both built *from*, and representations *of*, the spatial practices that engender change and unsettle the status quo, and are close in character to Massey's relational interdependence and 'throwntogetherness' of space (Massey, 2009, pp. 140–141). In contrast to assumptions of change coming from class revolution, this positive conception of difference as integral to the right to produce space is articulated from both ideas of spontaneity and festival, and the material reality of the everyday (Goonewardena et al., 2008a, p. 296; Lefebvre, 2009, p. 16).

Such spatial relations of difference question what is possible and impossible. They are not prescribed by representations of space and spaces of representation, but are born from the complexity of difference and lived space:

> The affirmation of difference can include (selectively, that is, during a critical check of their coherence and authenticity) ethnic, linguistic, local, and regional particularities, but on another level, one where differences are perceived and conceived as such; that is, through their relations and no longer in isolation as particularities. Inevitably, conflicts will arise between differences and particularities, just as there are conflicts between current interests and possibilities. Nonetheless, the urban can be defined as a place where differences know one another and, through their mutual recognition, test one another, and in this way are strengthened and weakened.
>
> (Lefebvre, 2003, p. 98)

It is this affirmation and valuing of difference that allows Lefebvre's spatial critique to transcend beyond a classical and intransigent political 'right to the city'. The 'right to difference' goes further. It reveals the somewhat overlooked spatial perspective of the contestation of difference as being intrinsic to the negotiation and confrontation of what is possible and impossible (Lefebvre, 1991, p. 60).

The right to the city contested as a right to potentially differential space is thus of great theoretical value to our study. It poses a radical re-contextualisation of the Marxist advocacy of change achieved through the political action of workers (Lefebvre, 2009, p. 120). In contrast Lefebvre proclaims that by remaking the city through the contestation of difference we can remake political and economic relations as part of a differential contestation of the spatial trialectic. By living differently and promoting different social relations we would engender representational spaces that respond and engage with such difference. Thus we might begin to conceive of development architecture as offering an alternative socio-spatial framework of space that supports grass-roots lived spaces of difference, rather than conceiving of its suppression in homogeneity (Lefebvre, 1991, p. 383). However, Liette Gilbert and Mustafa Dikec acutely surmise the implications and challenges of difference:

> Unless the forces of the free market, which dominate – and shape to a large extent – urban space, are modified, the right to the city would remain a seductive but impossible ideal for those who cannot bid for the dominated spaces of the city; those, in other words, who cannot freely exercise their right to the city.
>
> (Gilbert and Dikec, 2008, p. 261)

For Gilbert and Dikec the historical complacency of consensus and homogeneity is a daunting obstacle to social change. Yet what is of note is the particular reference given to how the right to difference does not simply engage with everyone living in cities equally, but questions the power geometries at play within cities and social space. The right to difference implies an engagement with people who cannot freely engage with exercise their right to the city. This is a far more challenging vehicle from which to engage with space, and in particular the contradictions and contrasts between the inequalities of space in the Global North and South.

As has been previously noted, this book is explicitly not a critique of Western space or of cities in the Global South. However, in comparison to Lefebvre's articulation of the positive potential of differential space we can explore and reveal alternative examples and methodologies of realising differential space. By critically comparing the theoretical social implications of differential space against the informal settlements of the Global South our analysis suggests provocative methodological insights into the appropriation and participation in space that can engender change and contestation in a non-Western context.

Thus we arrive at the contention that Lefebvre's concept of differential space is a critical lens through which to interpret the implications, possibilities, and positive potential of informal settlements in the Global South. And subsequently, our proposal to connect Lefebvre to Massey might provide a theoretical framework to globalise the question of the right to difference, and might also further reinforce the opportunity to re-imagine space and architecture by learning from other peoples, informal space, and alternative development practices from the Global South.

Informal spaces of development advocated by the likes of Turner and Hamdi are produced out of material, economic, political, and social necessity – out of a spontaneity and invention that necessitates a continual contestation of social relations and the production of provocative alternative, informal, and socially different spaces. Informal housing settlements, slums, squats, favelas, and barrios are perhaps the epitome of differential space:

> Differences endure or arise on the margins of the homogenised realm, either in the form of resistances or in the form of externalities (lateral heterotopical, meteorological). What is different is, to begin with, what is excluded: the edges of the city, shanty towns, the spaces of forbidden games, of guerrilla war, of war. Sooner or later, however, the existing centre and the forces of homogenisation must seek to absorb all such differences, and they all succeed if these retain a defensive posture and no counterattack is mounted on from their side. In the latter event, centrality and normality will be tested as to the limit of their power to integrate, to recuperate, to destroy whatever has transgressed. [...] The vast shanty towns of Latin America (favelas, barrios, ranchos) manifest a social life far more intense than the bourgeois districts of the cities. [...] The result – on the ground – is an extraordinary spatial quality.
> (Lefebvre, 1991, p. 373)

Originating from a confrontation of the inequality of Victorian industrial Manchester, a contemporary globalised contestation of 'the right to the city' engages the same issues but outside a singular Western spatial field of reference.[12] However the right to the city as an expression of difference in the Global South is more pronounced. The extreme polarisation of equality that prevails in informal settlements and economies of absence makes such rights necessities for survival (Lummis, 2010, pp. 38–54, 1997). The city in the context of necessity and difference is implicitly understood as both a right and a responsibility. When such rights do not exist they are, out of necessity, spontaneously seized, appropriated, and taken creating informal settlements and differential city spaces. This was the situation Turner encountered in 1960s Peru. The same situation exists today throughout cities in the Global South and increasingly in the Global North, necessitating the kind of critical research that this book begins to frame.

The underlying reasoning of this comparative analysis is marked here by an explicit re-contextualisation of interdisciplinary discourses in order to engage with the social and political potential of positive differential space. Thus the participatory methodologies that underpin the development methodologies of Turner and Hamdi can be seen as spatial practices that facilitate the gestation of such differential space. Provocatively such practices are not defined by conventional concrete architectural interventions, but are articulated through confrontation and continual contestation of the broader social and spatial relations of place (Massey, 2009, p. 19).

The right to difference

Through Lefebvre we recognise the potential of the city as a contestation of the political and social potential of understanding space as a product (Elden, 2004, p. 144). However our premise that the right to difference is a fundamental transition in Lefebvre's discourse prompts further exploration of the positive spatial implications of difference as a means to socially produce alternative relations of space. In other words, if the right to the city is merely the right to abstract space then there is no value in it as part of a social process, and space becomes inanimate, passive, and tame (Massey, 2005, p. 23). Here development practice, cities, and informal spaces of the Global South can begin to be re-imagined as spaces that reveal and highlight the hidden differences and contradictions of capitalist hegemonic relations. This connection of the city, difference, rights, and hegemony provides a foundation for the comparisons and trajectories explored in the development methodologies of Nabeel Hamdi in later chapters.

The right to difference is thus an overarching principal of positive Lefebvrian space that transcends the right to the city (Merrifield, 2006, p. 113). It implicates the inherent positive potential of Lefebvre's 'social space as a social product' as an explicit means to counter space as an abstract product (or commodity).

Lefebvre is clear that only through practical action is the social responsibility for space made manifest in its articulation and contestation of different possibilities of space through social relations (1991, p. 52). Contestation, appropriation, and transgressions articulate the potential difference and alternative relations of space to contest the inevitability of cultural hegemony. Thus Lefebvre's advocacy of the critical re-appropriation and re-articulation of social space as a social product is articulated towards the possibility of differential space, and the contestation of the conceived and perceived constructions of abstract space through the material reality and practices of the lived everyday (Lefebvre, 2009, pp. 202–205; Merrifield, 2006, p. 112).

Lefebvre's differential space implies the positive potential of liberation and change. Contrary to difference as merely that which rejects or counters the homogeneity of abstract space, Lefebvre explicitly advocates space as a 'geography of different rights' that moves beyond rights in general (1970, p. 45). The struggle to implement such rights implies a confrontation with the homogenising power of ideologies, centralities, and unlimited growth. These ideologies enforce themselves through supposed technical and scientific rationality of formal and abstract space, destroying the particular and differential possibilities (Lefebvre, 2003, p. 96). Yet here, this chapter is confronted with both the great insight and complication of Lefebvre's approach. He articulates the dialectical necessity for difference as a vehicle for social and political contestation, but always leaves the reader with an ambiguity and illusiveness of form as to how such spatial contestation might come to be more than temporary and ephemeral:

> With differential space, Lefebvre plays his Nietzschean-Marxist trump card at a decisive moment [...] Differential space isn't systematic, and so the form and content of *The Production of Space* unfolds eruptively and disruptively, unsystematically through a Nietzschean process of 'self-abnegation'. [...] Nothing even remotely resembles a system [...] neither in form nor in content. 'It's all a question of living,' he explained in the closing lines of Le manifeste differentialiste. 'Not just of thinking differently, but of being different.'
>
> (Merrifield, 2006, p. 17)

In the context of our research premise, such an advocacy of the positive potential of Lefebvre's 'being different' is comparable to the material or social scarcity that drives dialectic social change observed by Turner in informal settlements. Lefebvre's contention that difference must be fought for is perhaps revealing of the passive acceptance that is often observed as afflicting Westernised space – becoming mere spectacle for consumption (Debord, 1994). In contrast spaces, practices, and relations of difference in the Global South reflect a manifestation of change that is driven from the practical and grass-roots dialectical exploration of alterity, scarcity, and necessity. The implication that difference is found within the interdependence of scarcity, necessity, and freedom is suggestive of a re-alignment of the lived, conceived, and perceived trialectic in confrontation with political ideology:

> The reawakening of a 'politics of difference' (as opposed to the tendency of homogenisation), in which the rich creativity of the excluded can be

developed into a concrete alternative to the present spatial system. Lefebvre detected this in his Latin American travels and stays in the slums and favellas of Brazil, which appeared to be moments in which alternative local spatialisations were brought into existence. Was he a naive romantic?

(Shields, 1999, p. 183)

This chapter and wider book contend that by looking beyond the confines of Western cities and space we can learn from spatial practices that challenged abstracted hierarchies and social hegemonic trajectories by the simple necessity of being different; we can learn from the development contexts of informal space in the Global South. This global contextualisation of Lefebvre's differential space-time provides an explicit intersection with Massey's advocacy of the positive interpretation of relational and interdependent space as a complex multiplicity (2005, pp. 180–183). For Massey, both formal and informal types of urban space are inherently interdependent and relationally constructed (Harcourt et al., 2013, p. 170). Yet local and global differences of culture, economy, and social relations become an opportunity to recognise, contest, and negotiate the material dialectic practices that produce difference. If we interpret space in this way then differences cannot be isolated. They exist as active connections that allow such relations of difference to resist structural antagonism and dichotomy (Goonewardena et al., 2008a, p. 296).

Where might we look for examples of differential space and the right to difference? In line with our connections with spaces of the Global South we could reasonably cite Haas and Hahn's favela painting projects in Rio de Janeiro. Whilst these simple projects might not reflect the full complexity of differential space they do aspire towards the same production and valuing of alternative spaces and social relations. The simple act of painting favela buildings that are conventionally (from a Westernised perspective) deemed to be temporary, degenerate, or failing posits a positive acceptance and engagement with strengthening, improving, and celebrating informality and difference.

Yet the favela example is seemingly far removed from the potential of differential space in a Westernised context. Looking closer to home we might rely upon Justin McGuirk – who is perhaps the most recognisable recent proponent of learning from alternative spaces and/or the Global South (2014) – and, for example, his recent advocacy of a contemporary 'Urban Commons' (2015). For example McGuirk cites Atelier d'Architecture Autogérée and Doina Petrescou's Urban Commons in Colombes, Paris. Using these examples he explicitly highlights the positivity of this Paris community's achievement in transitioning from simple community gardening to cooperative housing and recycling. Scepticism about the potential broad replicability of this project aside, this is a differential space – a positive engagement with alternative spaces and social relations. These projects must be recognised and valued, but must also be critically engaged with to discover how they can become more than mere occasional occurrences and develop instead into regular aspects of Westernised space and life.

In contrast to engagements with formal and abstract space, interactions within and between differential spaces and alternative social relations inherently produce social interactions, positive spatial agonism, and the potential of eventual relational mutuality (Massey, 1994, pp. 265–266). This positive conception of difference advocates embracing the potential of the inevitable conflicts, agonisms, and intersections of differences that transcend the cohesion of abstract spatial relations and Westernised spaces of neoliberal capitalist hegemony (Kipfer, 2008, p. 206):

> Such alternative and oppositional claims for difference can take on very different forms and ways of expression: small-scale remittances, counter-projects, anti-imperial insurgencies, rebellions of the dispossessed in metropolitan centres such as the recent uprisings in Paris, as well as well-documented anti-globalisation struggles and networked encounters. Struggles of peripheralised social groups against segregation and for empowerment can produce their own forms of centrality. [...] The search for new centralities in a contest of translational urbanisation thus leads not only to global and capital cities (New York and London) but also to central places produced by counter-networks and mobilisations (Porto Alegre and Bamako).
>
> (Goonewardena et al., 2008a, p. 296)

Here our analysis aligns with the underlying post-structural and post-colonial tenets of Kanisha Goonewardena et al.'s recent text *Space, Difference, Everyday Life: Reading Henri Lefebvre* (2008b). In seeking to transcend the economic and post-modern analyses of Lefebvre, Goonewardena et al. propose a 'third reading' of Lefebvre with the intention of exploring 'a heterodox and open-ended historical materialism that is committed to an embodied, passionately engaged, and politically charged form of critical knowledge' (2008b, p. 2).[13]

This third reading of Lefebvre offers a first critical intersection with Homi Bhabha's 'third-spaces' and 'cultural hybridity' (2004, pp. 53–56). In connection with Lefebvre's differential space this re-reading contextualises cultural identity and difference within the anti-representational implications of post-colonial space: 'radical openness,' 'otherness,' 'margins', and 'hybridity,' where 'everything comes together' in a place of 'all inclusive simultaneity' (Goonewardena et al., 2008b, p. 9). This conception of space as both open-ended[14] and interdependent with lived everyday space can be seen vividly in Massey's advocacy of space as 'the sphere of coexisting heterogeneity' and of a 'simultaneity of stories so far' (2005, p. 24). Similarly for Lefebvre, to re-assemble the positive potential of such differential lived space required the social transformation of fragments or moments or rhythms of positive difference in urban life, not its coalescence and reification as within the structural abstractions and reification of modernity.

For Goonewardena et al., the post-structural appropriations of Lefebvre led by Bhabha and similarly the urban political-economic renderings of

Harvey largely fail in their attempt to 'overcome the divide between culturalism and economism in a substantive way' (2008b, p. 9). Our analysis of development practices would contend a similar critique, but articulated towards the lack of a critical dichotomy between both theoretical and practical discourses on space. Goonewardena's critique provides a provocative theoretical intersection with our wider goal of contextualising Lefebvre in the global, plural, and positive spatial context of difference. Yet this attempt at a critical expansion of Lefebvre's exploration of difference remains conflicted in its critique and rejection of many potential post-structural adaptations of Lefebvre.

Of particular interest is Kipfer's rejection of Derrida's post-structural *différance* as a viable mechanism to contest the cultural practice and identity of differential space (2008, p. 202). This would seem to suggest that the necessary pluralism of spaces that is contested in cultural difference and differential space is an underlying problem that even the provocative insights of Andrew Schmuely (2008) and Richard Milgrom (2008) respectively cannot quite transcend. This missed opportunity in avoiding Derrida's post-structural articulation of difference is remarkably similar to Massey's critique of the inherent negativity of deconstructive *othering* (2005, p. 93). Yet in contrast Massey actually extracts crucial value and positive potential from Derrida that we will explore in detail in later chapters.

In spite of this critique, Goonewardena et al. provide some of the clearest discourse with which to articulate the global implications and positive potential of 'the right to difference'. Their work is thus utilised here in an attempt to generate a wider examination of the urban question that transcends the structural dichotomies and traditional mechanisms of political contestation through organised protest and antagonism:

> To this end, three tasks will be vital. First, it is important to grasp the basic construction of Lefebvre's epistemology in order to achieve a sound theoretical basis for empirical analysis. Second, fruitful applications of Lefebvre's theory have to be found. Manifold possibilities have arisen for this purpose, which remain to be fully explored. Some promising analyses do exist, however. Third, the crucial point of Lefebvre's approach should be taken into consideration: to go beyond philosophy and theory, and to arrive at practice and action.
>
> (Schmid, 2008, p. 43)

The trajectory of this chapter and wider book is itself intended as such a 'fruitful appropriation and application' of Lefebvre's theory, explicitly seeking to provide practicable and realised examples of positive and alternative social production of space. Here the exhortation 'to go beyond philosophy and theory, and to arrive at practice and action' once again is powerfully evoked in the comparative analysis that we are drawing with development practice methodologies. One of the great unspoken implications of Lefebvrian critique remains that such examples of practical realisations of social spatial relations are incredibly difficult to discern in a Western context. Even more difficult is to derive fruitful spatial methodologies that underpin such spaces in a positive and meaningful way. It is here that the interconnected comparisons observed in our analysis offer new additions to discourse concerning what the concrete realisation of Lefebvrian space might imply for spatial practices, the agency of architecture, and its methodologies of participation.

Appropriation and difference

In the context of the ideas and connections already outlined in this chapter it is necessary here to explore further intersections of Lefebvre's and Massey's respective discourses by connecting the spatial concepts of appropriation and difference. Many contemporary examinations of the positive potential of appropriating and re-appropriating 'the city' rely upon the concept of appropriation as a spatial process (see for example: Garrett, 2013; Harvey, 2012). Lefebvre's discourses on spontaneity, contestation, and appropriation are explored throughout various texts, notably *The Explosion* (1969) and *The Production of Space* (1991). Yet the practical reality of spatial appropriation and its connection to the right to difference remains problematic. As noted by Walter Prigge, 'if social power is symbolised in the appropriation of space, the significance of such spatialisation is revealed only through an analysis of these relations of meaning' (2008, p. 48). In somewhat of a reflection of this complexity of relational space, Lefebvre's interpretation of appropriation remains problematically balanced between the notion of the city as 'an exquisite oeuvre of practice and civilisation' (1996, p. 126), and the explicit political articulation of contestation and spontaneity as a refusal to be integrated.

This complex articulation of the positive political potential of the social appropriation of space is apparently reliant upon both spontaneous moments of everyday festival, and a political counter-revolutionary uprising from the streets. Protest and revolutionary uprising were at the forefront of Lefebvre's writing in *The Explosion* and other works due to his explicit role and engagement in the 1968 Paris student protests.

Thus, the city as oeuvre 'is use value and the product is exchange value. The eminent use of the city, that is, of its streets and squares, buildings and monuments, is la fete' (Lefebvre, 1996, p. 66); whereas in contrast, contestation is born from negation and is articulated by the negative characteristics it brings to light from its place of origin: 'it surges from the depths to the political summits, which it also illuminates in rejecting them' (Lefebvre, 1969, p. 67).

Thus whilst the festival defines a moment in which the world is turned upside down, re-imagined and only symbolically re-enacted, revolution is different – it is for real. It is this disjunctive contrast between Lefebvre's propositions for both positive and negative articulations of appropriation that defines his theoretical and political critique of use value and exchange value. Trapped in the contradiction between the festival of the everyday and the Marxist notion of revolutionary upheaval through class struggle, Lefebvre's critique remains focused upon the danger of the co-option of differential spatial relations rather than a practical positive advocacy (1991, p. 356).

Thus appropriation and the contestation of difference is a rallying cry for the urban populous to become apart of social production of space. Yet such theoretical and ideological appropriations run the risk of overlooking the quotidian necessity of Lefebvre's spatial, material, and dialectical propositions (2008, p. 676). As a consequence of these complex academic articulations of appropriation, practical applications of Lefebvre's observations and advocacy of the significance of everyday life as a site of spontaneity, invention, and appropriation remain linked to the potential for a working-class revolution in the (predominantly Westernised) city (Merrifield, 2006, p. 92).

In contrast, a clearer articulation of what this book seeks to explore exists in the connection between the contestation that arises in 'the right to difference', and Lefebvre's proposition of appropriation as interdependent with quotidian and spontaneous moments. This more sensory and material articulation of difference and appropriation as part of the everyday suggests a more practicable and grass-roots political engagement with the social relations that articulate the production of space. It is in the context of this practical and material articulation of appropriation as a positive social agonism[15] of difference and multiplicity that Lefebvre's social practices of appropriation and change can be critically compared to the grass-roots social change of alternative participatory development (2009, p. 150). Thus Lefebvre frees contestation from mere negativity and revolutionary ideology and offers the potential for positive contestation and struggle in concrete material problems.

> The encounter brings politics back into the city by breaking the circuit of endless reproduction, of ideology masquerading as politics. It becomes a short circuit in a web of social relations. The city itself becomes the privileged subject/object, rather than mere location, of philosophy: the perception of the city as form, as an expression of 'situated knowledge' (the phrase of Bahktin's), constitutes an aesthetic praxis. [...] Lefebvre recovers the utopian potential of aesthetic mediation as a privileged expression of appropriations of the spatio-temporal.
>
> (Nadal-Meslio, 2008, p. 167)

With the explicit articulation that it is not enough to simply produce difference and that it is necessary to see difference as part of lived experience and praxis, Lefebvre's discourse of appropriation with spontaneity and moment can be perceived as recognising an explicit value in the social contestation of the material reality of the everyday (Schmid, 2008, p. 28). This everyday contestation as revolution reflects Massey's advocacy of the daily negotiation and contestation of place (2004, p. 11), and similarly Jean-Luc Nancy's identification of political space as 'a community consciously undergoing the experience of its sharing' (1991, p. 40). Through their shared articulation and value of spatial and social difference Massey and Lefebvre can be critically observed as intersecting within the identification of place as formed 'through a myriad of practices of quotidian negotiation and contestation; practices, moreover, through which the constituent "identities" are also themselves continually moulded' (Massey, 2005, p. 154).

This interpretation of Lefebvre's appropriation as a relationally inter-dependent contestation of the everyday re-frames his earlier discourse on the implications of materials context, lived spaces, and the social relations of production (Shields, 1999, p. 119). As exemplified in our previous comparison of dialectical materialism with Turner's 'housing as a verb', this relational interdependence is articulated through a model of informal appropriation that produces social relations through participation, action, and a more positive identification of the contestation of space:

> Central to Lefebvre's materialist theory are human beings in their corporeality and sensuousness, with their sensitivity and imagination, their thinking and their ideologies; human beings who enter into relationships with each other through their activity and practice.
>
> (Schmid, 2008, p. 28)

In the global context of our comparisons with development practice, this alternative articulation of activity and practice contests the reification and abstraction of space in Western cities and society in general (Lefebvre, 2009, pp. 187–193). Western architectural discourse has repeatedly engaged with such alternative methodologies of spatial appropriation and difference (Mason, 2012). We could cite Iain Borden's discourse on skateboarding (2001), or Adam Evans discussion of graffiti and Hip Hop culture as spatial engagements with culture, appropriation, and alterity (2014). Yet Bourdieu's articulation of cultural capital and the reification of such 'alternative' practices reflect the potential limitations of spatial appropriations that agonistically contrast conventional capitalist consumption (2010, pp. 3–6, 1991, p. 113).

Thus we must resist the temptation to suggest that appropriation only takes place in times of revolution and not in simple everyday lived spaces. Both may equally express positive spatial differences, yet both can be critiqued as co-opted and pacified by the logical cohesion, security, and hegemonic comfort of capitalist space (Lefebvre, 2009, p. 69, 2003, p. 187). This reification and capitalist absorption of spatial difference and appropriation abounds in Westernised space, but is far less apparent in the emergent informal spaces of the Global South. Robert Neuwirth notes the mis-identification of informal spaces and economies as fragile or inefficient instead of recognising them as dynamic and responsive engagements with the positive potential of informal spatial relations and socially sustainable development (2012). These spaces might thus stand as practical spatial realisations of Lefebvre's confrontation of the imbalance of the conceived and perceived over the material reality of the lived:

> The practice of appropriation [...] manifests a higher, more complex rationality than the abstract rationality of modernism. Significantly, as in

Hegel's category of concrete universal, these steps from the abstract to the concrete are seen as a sequence of differentiations: Lefebvre writes explicitly that the inhabitants produce differences in an undifferentiated space.

(Stanek, 2008, p. 66)

Appropriation and the alternative social relations that articulate the production of differential space offer a clearer articulation of what we might aspire to in the contestation that arises in 'the right to difference'. By proposing appropriation as an articulation of social practices that are relationally constructed in informal spaces and particularly in the development spaces of the Global South, the potential for positive contestation and struggle is renewed in concrete material problems of globalised inequality instead of classical visions of political and class based ideologies. Appropriation was part of a social process Lefebvre would label 'cultivated spontaneity' in *The Survival of Capitalism*.[16] Yet for Lefebvre, contestation and struggle, transgression, and creation are dialectically interdependent:

Transgression without prior project, pursues its work. It leaps over boundaries, liberates, wipes out limits and advocates the far more explicitly political necessity of an 'explosion of unfettered speech'.

(Lefebvre, 1976, pp. 118–119)

In this context the question remains of how to cultivate such spontaneity without providing the formal fixity of the prior project, and particularly in light of the abstract technical and bureaucratic space, of modern urban life (Merrifield, 2006, p. 66). This chapter seeks to articulate and reinforce our underlying comparison of the positive aspirations of this explicitly Western spatial theory with the participatory methodologies of development practice that will be explored in later chapters. Thus the trajectory of theoretical spatial discourse connecting Lefebvre's differential space to Massey's relational multiplicity provides an explicitly positive articulation of appropriation, participation and contestation (Massey, 2005, p. 10). Complementary dialogues of the material dialectics, inequality, and contradictory space of participatory development practice can thus begin to be re-contextualised as concrete realisations of the positive potential of informal appropriation, difference, and multiplicity:

How can this homogenising 'contradictory space' become a differential space that particularises and humanises? Against conflict approaches, which begin with the assumption of the primacy of conflict in the relations between economic groups as the basis for the study of society, Lefebvre's formulation poses the disturbing question of people's cooperation, docility and complicitous self-implication on systems of inequality. [...] In answer to these issues, the spatial problematic draws attention to the symbolic and distorted forms of resistance practised through the spatialisation itself: eruptions of instability through the carefully spread net of Cartesian three-dimensional grid of rational and homogeneous modernity. Space itself becomes at once the medium of compliance and resistance.

(Shields, 1999, p. 183)

Participation and difference

Our analysis and comparison of Lefebvre's and Massey's discourse highlights a critical intersection of appropriation and spontaneity as spatial practices that are articulated towards the value of the positive contestation of difference. This connection is reinforced in our comparison of the intersection of differential space with participatory practice, and our critical re-reading of Lefebvre's interpretation of social transgressions by placing them in comparison with intersecting theoretical discourses of dialogue (Bakhtin, 1999; Koczanowicz, 2011), alterity, and multiplicity (Mouffe, 2013, pp. 22–23). These themes are explored variously in later chapters, however they are introduced here in order to expand the relation of contestation and appropriation to connect with concepts of participation and difference. Thus it is compelling to find Sara Nadal-Meslio making explicit references to action, spontaneity, and the 'Lefebvrian event' as a spatio-temporal act of social relation and participation:

> The spontaneity of the desire to connect is unequivocally political and has much in common with the lesser known Bahktinian theorisations of the 'act' in his unfinished *Towards a Philosophy of the Act*. In Bakhtin's words the act 'brings together the sense and the fact, the universal and the individual, the real and the ideal.' As we have seen, the revolorisation of the everyday, enacted through the aesthetic, as the natural milieu of both the 'event' and the 'act,' as a site for the enactment of being as event, is the prerequisite for both the 'act' and the 'event' to explode.
>
> (Nadal-Meslio, 2008, p. 169)

Within Lefebvre's conception of appropriation and contestation as interdependent there remains an implicit connection with the notion of spatial participation (Lefebvre, 1969, p. 68). However his articulations of appropriation are also interconnected with the political representation and manipulation of space, revealing that the notion of participation can also be negatively implicated in the co-option and oppression of abstract formal and bureaucratic space. Here Lefebvre employs his spatial articulation of autogestion, as explored in the previous chapter, as a counter to this potential co-option of participation as merely a mechanism for placating and quieting the populous (2009, p. 150, 1969, p. 84). However the questions posed by our comparison to both Turner and Massey contend that Lefebvre's advocacy of autogestion can be critically compared with post-structural participation discourses and development practice to offer novel methodologies to contend and confront the potential co-option of spontaneity, difference, and participation.

In their exploration of practical methodologies of participation, development discourses such as *The Development Dictionary* (Sachs, 2010) and *Participation: The New Tyranny?* (Cooke and Kothari, 2001) specifically contest the same spatial questions of difference, multiplicity, and positive appropriation that our comparisons have extracted from Lefebvre and Massey. Reflecting Lefebvre's analysis of the dichotomies of participation and difference, co-option and autogestion, the re-contextualisation and comparison of informal settlements and development practice similarly implicates participation as interdependent with both the positive potential of multiplicity and negative implications of co-option and ideology (Rahnema, 2010, p. 132).

In *Participation: From Tyranny to Transformation* (2004) Sam Hickey and Giles Mohan offer a new critique of the inevitability of global co-option, ideology, and inevitability in participatory development, posing the positive potential of facilitated (or cultivated) participation. This critical and challenging examination of the political complexity of development space and global policy is explored in more detail in later chapters, however its intersection here with Lefebvre's positive potential of differential space is provocatively accurate:

> The idea of 'togetherness in difference' is based on the interspersion and interaction of difference theories. While differences exist, there is also the recognition that relational identities require multiple others so that the identity of one depends upon other/s, which gives groups a mutual stake in one another's existence. At various levels this opens up the possibility that alliances exist since only some differences are intractable.
>
> (Hickey and Mohan, 2004b, p. 64)

Andrea Cornwall's *Space of Transformation* (2004) explores a further explicit intersection of these trajectories of participation, difference, and appropriation. In connection with participatory development practice she frames a comparison of Lefebvrian spatial analysis of contestation, resistance, and agency against what she terms 'invited space of participation' (2004, p. 76). In contrast to strict structural boundaries of development discourse and theory, this inter-disciplinary comparison contests theoretically provocative explorations of participation and development as a spatial practice:

> Viewing participation as a spatial practice helps draw attention to the productive possibilities of power as well as the negative effects. [...] Social relations, Lefebvre contends, exist only in and through space; they have no reality outside the sites in which they are lived, experienced and practiced. And every space has its own history, and is threaded through with the traces of other histories, in other spaces, its own 'generative past'. [...] Spaces come to be defined by those who are invited into them, as well as by those doing the inviting.
>
> (Cornwall, 2004, p. 80)

This critique of the invited nature of participatory spaces articulates a theoretical discussion of the complex relations of political, economic, and spatial power that practical development methodologies are engaged, explicitly referencing Lefebvre's conception of the trialectic interdependence of space, power, and representation. The explicit confrontation of the relations of power and agency is recognised in the significance of the distinction (and difference) between those being invited to participate, and those with the power to invite (Cornwall, 2004, p. 76).

In order for participation and change to be transformative and to resist this effect of co-option, alienation, and reification, they cannot be defined by invited spaces. Instead they must become an interdependent part of the everyday social relations, self-management, and participatory production of space (Lefebvre, 1969, p. 90). Here participation and appropriation are co-implicated in an articulation of positive difference and heterogeneity and the consequences and challenges of this premise are not lost on Cornwall:

> Spaces in which citizens are invited to participate, as well as those they create for themselves, are never neutral. Infused with existing relations if power, interactions within them may come to reproduce rather than challenge hierarchies and inequalities. Yet the 'strategic reversibility' of power relations means that such governmental practices and 'regimes of truth' in themselves are always sites of resistance; they produce possibilities for subversion, appropriation and reconstitution.
>
> (Cornwall, 2004, p. 81)

In her utilisation of Lefebvre's spatial discourse Cornwall does not abandon the positive potential of invited spaces of participation. Instead she continues to pursue the socially transformative potential of development through positive and critical spatial practices, observing that the comparison of participation as a spatial practice in a Lefebvrian framework offers a particularly useful critical lens of analysis (2002). In this comparison, spaces of participation can thus be critiqued in terms of the situated nature of such practice within 'bounded yet permeable arenas in which participation is invited, and the domains from within which new intermediary institutions and opportunities for citizen involvement have been fashioned' (Cornwall, 2004, p. 75).

This analysis proposes participation as both a positive and a situated spatial practice of socio-political contestation of differences and social relations of inequality. Thus when considered in comparison with Lefebvre's differential appropriations of space, Cornwall's articulation of participation can potentially confront and contest the dynamics of power, voice, and agency, revealing their material and spatial qualities in order to facilitate the necessary spatial and strategic turns towards a 'more genuinely transformative social action' (2004, p. 75, 2002, p. 75).

In Western space the connections between community participation, space, and architecture appear to once again exist only in the (academic/intellectualised) periphery of social space. In a neoliberal political and economic context, participation is reduced to the insipid placation and mis-direction of local communities as part of contrived planning processes, or to explicitly direct social enterprise undertaken by the liberal elite of architectural practice. Please note I do not seek to belittle the achievements of Western architects, but merely seek to question, in much the same way as does Markus Miessen (2007, 2011), the long-term economic and social sustainability of such practices.

Thus, Cornwall's question of the power relations of such invited spaces of participation wreaks even further havoc on such discourse. In Western space the invitation to participate exists as either a passive agent of formal planning bureaucracy, or is limited to the achievements of an architectural liberal elite. The long-term efficacy of any such invitations to participate seemingly lack plausible sustainability, and perhaps even limit the potential for the necessary social participation as part of everyday life.

Yet the conception of invited spaces in participation remain for Cornwall a viable positive medium for social change precisely because of the relational potential of space she draws from Lefebvre's discourse. In the context of both Lefebvre's and Massey's articulations of the positive potential of space, this crucial connection to the agency of participation confronts, contests, and appropriates the potential of differential and relational space through the spatial practice of participation. Considered in this way, participation is integral to the production and social practice of space as a political forum and theatre through which positive social debate and transformation can be performed and cultivated:

> 'invited spaces' bring together, almost by definition, a very heterogenous set of actors among whom there might be expected to be significant differences in status. [...] 'invited spaces' assemble people who might relate very differently if they met in other settings, who may be seen (even if they don't see themselves) as representing particular interests, and who generally have rather different stakes in, accountabilities for and responsibilities following any given outcome.
>
> (Cornwall, 2004, p. 76)

Thus in comparison with Lefebvre's spatial practices Cornwall's positive notion of invited space exists as a forum in which to assemble heterogenous actors and agency. It is an articulation of participation that intersects with spatial appropriation and the facilitation of 'the right to difference'. Similarly, the dialogue and negotiation that is implied and necessitated by invited spaces of participation articulates the political possibility of the negotiation of sociospatial differences. The act of negotiation provides the underlying positive contestation of social relations as a spatial practice of dialogue, negotiation, and agonistic contestation, moving towards the social production of alternative social space (Habermas, 1984, pp. 80–81). As a consequence of this analysis there remains for Cornwall an explicit awareness of the social and political implications of identity, marginalisation, and authority that are entailed within spaces which are 'invited' rather than spontaneous:

> The contrast here between spaces that are chosen, fashioned and claimed by those at the margins – those sites of radical possibility – and spaces into which those who are considered marginal are invited, resonates with some of the paradoxes of participation in development. Yet the boundaries between such spaces are unstable: those who participate in any given space are also, necessarily, participants in others; moving between domains of association, people carry with them experiences and expectations that influence how they make use of their agency when they are invited to participate, or when they create their own spaces. And the scope that 'invited' or 'popular' spaces offer for political agency is, in turn, influenced by a host of contextual factors. Analysed through the lens of the concept of space, the political ambiguities of participation become all the more evident.
>
> (Cornwall, 2004, p. 78)

Exploring this issue of the co-option of participation further Cornwall explicitly utilises Lefebvre's conception of space as 'not simply "there", a neutral container waiting to be filled, but as a dynamic, humanly constructed means of control and hence of domination, of power' (1991, p. 24). Viewing participation as a spatial practice helps to draw attention to inherent productive possibilities of power as well as the more frequently focused-upon negative implications. Here Cornwall's analysis and discourse aligns once again with Massey's advocacy of the positive multiplicity of space (Massey, 2005, p. 100). If social relations are produced and reproduced only through the specificity and relationality of social sites then every place has its own history and truly relational places exist as a spatial practice, and as such express traces of their own stories and social 'generative past' (Lefebvre, 2009, p. 100). Thus for Cornwall: 'Spaces come to be defined by those who are invited into them, as well as by those doing the inviting.' (2004, p. 80). These socio-spatial relations define the traces and stories of place and identity whether for good or bad or all the complex grey areas in-between.

This interjection of participation as a politically agonistic and positive Lefebvrian spatial practice culminates with Cornwall's utilisation of James C. Scott's original discourse on the practices of informal spaces and development (see: Scott, 1987, 1990, 1998). This series of intersecting comparisons begin to reinforce a confirmation of the interdisciplinary and comparative methodology used in our analysis, and the necessity to reveal and contest the hidden potential relationships between the theoretical and practical observations of socially, spatially, and relationally produced space:

> Scott's [...] explicit concern with the spatiality of power and resistance offers useful analytic tools for making sense of the shaping of spaces, and for exploring the potential of differently located spaces. [...] Exploring the

extent to which such 'weapons of the weak' are deployed on spaces for participation may be instructive: agendas can be shaped as much through pretending not to understand, remaining silent, staging an argument, taking all at once, as by articulating positions openly.

(Cornwall, 2004, p. 82)

The provocative connection of participation and political spatial practice suggests an interdependence of the politically ambiguous potential of participation and the complexity and contradictions of Lefebvre's spatial trialectic. Within this critical framework, the complex social and political ambiguities of the various actors in participatory development that Turner and Hamdi engage with in their practices reflect both Lefebvre's interdependent conceived, perceived, and lived spaces (Shields, 1999, p. 120), and Massey's conception of space as the 'sphere of positive multiplicities' (Dikec and Dikeç, 2012, p. 673).

The political and spatial implications and positive potential of participation methodologies is explored in much greater detail in later chapters. However, our comparative re-reading of Lefebvre's and Massey's respective articulations of the positive potential of difference and multiplicity has been crucial in describing a trajectory of spatial discourse that intersects with development discourse and the contemporary critical question of participation. Here we find resonance with the Marxist critiques of the co-option of participation and appropriation that are possible when Turner's discourse is mis-interpreted and oversimplified to simple dichotomies of sweat-equity and self-help. As explored previously, by re-reading examples from practitioners like Turner, participation can be revealed and potentially re-appropriated as a contestation of political freedom, autonomy, and choice. The potential of positive participation like this reveals a renewed conviction of practical space as the medium for change that is made manifest in spatial relations and practices as an ongoing and unfolding social process.

Positive multiplicity

In our reading of Lefebvre's articulation of differential space and appropriation the connection to the global context of critical participatory development practice is invaluable to our wider premise of re-imagining architecture and space. Having explored this critical trajectory, it is now possible to offer further critical connections to Massey's discourse by contextualising the key spatial concept, multiplicity. This re-contextualisation and analysis offers a critical framework from which to perceive new comparisons and intersections of space to questions of geography, politics, and the economics of global development.

Massey's global and relational critique of structuralism's negative and passive interpretations of space emanates primarily from a geographical interrogation of modern structural systems of differentiation and organisation into bounded places (Featherstone and Korf, 2012, pp. 663–665). Her discourse contests the implications of structuralism for conceptions, representations, and lived experiences of space specifically by confronting how spatial difference has been systematically convened into temporal sequence. Or to paraphrase, how development has become structurally tamed within an ideological progression towards a singular Westernised prescription of modernity (Bauman, 2000, pp. 59, 123).

Massey's relational interpretation of the production of space also reflects an almost identical articulation of space to that of Lefebvre's spatial triad. Her confrontation and contestation of abstract space and inevitable capitalist development intersects with Lefebvre's, Turner's, and Hamdi's various advocacies for spatial positivity, heterogeneity, and practice. These intersecting advocacies for positive differential space, and for a multiplicity of unfolding development trajectories, provide the basis for our wider post-structural interdisciplinary trajectories. In essence, we can take Lefebvre's right to difference and Massey's positive and relational multiplicity, and combine them to validate a re-imagination of Westernised architecture and space in critical comparison with examples of alternative, sustainable, and participatory development from the Global South.

Massey's analysis observes how different stories and lives are identified and organised within modernity and Westernised development as merely moments of existence within the sequential production and performance of a prescribed homogenous development (2005, p. 68). This critique of structural space, globalisation, and development highlights the socio-political implications of interpreting different places as merely different stages along a single linear temporal development. As a counter to this, Massey advocates an open, plural, and mutually interdependent positive space as relationally produced and implicated:

> For to open up 'space' to this kind of imagination means thinking about time and space as mutually imbricated and thinking of them as the product of interrelations. You can't go back in space-time. To think that you can is to deprive others of their ongoing independent stories. It may be 'going back home', or just imagining regions and countries as backward, as needing to catch up, or just taking that holiday in some 'unspoilt, timeless' spot. The point is the same. You can't go back […] You can't hold places still. What you can do is meet up with others, catch up with where another's history has got to 'now', but where that 'now' (more rigorously, that 'here and now', that *hic et nunc*) is itself constituted by nothing more than – precisely – that meeting up (again).
>
> (Massey, 2005, p. 124)

This globalised comparison of differential space and multiplicity marks a pivotal intersection in our theoretical framing of a re-imagination and articulation of the positive potential of space. In simple terms, our theoretical framework and critical lens posits that Massey's relational multiplicity of space can be re-read as a global articulation for Lefebvre's differential space.

Thus in critiquing the relationship between space and time to global development Massey's observation of spatial taming into a temporal sequence of
singular homogenous capitalist development provides an explicit lineage from
Lefebvre's spatial articulation of Marxist critical theory (Massey, 2005, pp. 4,
82). By introducing the complexity, implications, and potential of coeval
multiplicity (Fabian, 2002, p. 59), Lefebvre's observations of the social relations
and production space can be contested as a positive and plural conception of
difference and alterity in contemporary development practice (Goonewardena
et al., 2008a, p. 297; Lefebvre, 2009, p. 119).

As noted tangentially by Baldwin (2012), this comparison contests that
Massey's articulations of space and positive multiplicity offers an inherently
globalised yet locally interdependent interpretation of Lefebvre's exploration
of the production of space. Here the material, relational, and dialectic recognition of global inequality in development can be perceived as an expression
of the co-option of difference and the denial of multiplicity. It is a critique of
both global and local relations and the specificity of place, and of political
and economic practices (Massey, 2005, p. 173). It is also an explicit critique of
post-structural spatial discourses that critique space without challenging the
linear Western hegemony and subjectivity from which they come:

> Produced through and embedded in practices, from quotidian negotia-
> tions to global strategising, these implicit engagements of space feed
> back into and sustain wider understandings of the world. The trajec-
> tories of others can be immobilised while we proceed with our own; the
> real challenge of the contemporaneity of others can be deflected by their
> relegation to a past (backward, old-fashioned, archaic); the defensive
> enclosures of an essentialised place seem to enable a wider disengage-
> ment, and to provide a secure foundation. In that sense, each of the
> earlier ruminations provides an example of some kind of failure (delib-
> erate or not) of spatial imagination. Failure in the sense of being
> inadequate to face up to the challenges of space; a failure to take on
> board its coeval multiplicities, to accept its radical contemporaneity, to
> deal with its constitutive complexity.
>
> (Massey, 2005, p. 8)

Massey's introduction and utilisation of multiplicity reflects a desire and
necessity to recognise the positive potential of understanding space and time as
interdependent in place, and consequentially to engage with the plural and
open-ended potential difference in global development (2009, pp. 24–25, 2005,
p. 11). In contrast to the authority and quantifiable assumptions of structuralist
space the concept of multiplicity provokes a spatial counter-proposition to the
inequality of globalised Western hegemony and homogenisation (Hall et al.,
2013; Lefebvre, 2009, pp. 221–222).

In this context, Lefebvre's discourse on difference and appropriation is
implicitly a contestation of the same Western spatial reification and hegemonic

inequalities. Yet the abstract Western context of his analysis lacks the element of space as global multiplicity; it lacked the ability to contest differential space in a global sphere or to engage with the conception of planetary urbanism. Whilst the rapid acceleration of globalisation might explain this gap in Lefebvre's discourse, it is also noted that he was aware of global inequality and differential spaces (Merrifield, 2006, p. 73), thus suggesting a form of Eurocentric academic authority that was not contested till the advent of post-colonial theory (Chakrabarty, 2007).

A globalised and post-colonial world that is inherently defined by its multiplicity might imply the positive potential of pluralism and difference that Lefebvre strived to release from the historical inevitability of socialist politics and class revolutionary struggle (Lefebvre, 2009, p. 16; Shields, 1999, p. 125). Yet where appropriation, spontaneity, and contestation have lacked traction in the increasingly affluent and politically passive capitalist neoliberal contexts, Massey offers something different. Her notion of the global relationality and implications of space defines the positive political implications of space as a complex multiplicity. This multiplicity is not found or made, but exists as a natural and inherent part of space – the only choice is whether and how we perceive and engage with it. In the globalised context of both Massey's discourse and this book, multiplicity is an explicit confrontation with the interdependence of space, development, and inequality.

Positive pluralism and open-ended relational dialectic space imply the existence of a simultaneous multiplicity of lived world spaces – 'cross-cutting, intersecting, aligning with one another, or existing in relations of paradox or antagonism' (Massey, 2005, p. 3). Massey's social 'relations of paradox and antagonism' offer a Lefebvrian interpretation of the paradox of appropriation and difference with passive autogestion of social change (Lefebvre, 2009, pp. 138–146). Yet more provocatively this interpretation of spontaneity, difference, and alterity can also be recognised in our observations and analysis of Turner and Hamdi. Their examples and methodologies of participatory development highlight the positive implications of alternative and agonistic spatial practices in the socially sustainable development of informal settlements. Just as with Turner's advocacy of the autonomy and user-choice of informal housing, multiplicity implies that the relational specificity of place is key to empowering, and facilitating grass-roots change. Thus Massey's articulation of multiplicity is vital in our attempts to re-imagine Western space and architecture as it allows space to be recognised as an open and relational medium in which to explore, engage, and interact with positive heterogeneity of informal space (Massey, 2004, p. 13).

This positive and relational interpretation of multiplicity and space is itself somewhat complicated to conceive, especially when seen from the perceptive of Western abstract space – a space largely devoid of contestation and assured of its pre-eminence as the logical pinnacle of development (Bauman, 2000, pp. 69–72; Massey, 2004, p. 11, 1994, pp. 86–87). This is the challenge that Massey's advocacy of the multiplicity of space faces and it suggests the same potential for theoretical abstraction that Lefebvre posited in the co-option of social space and participation.

Yet when critically compared against Massey's discourse of multiplicity and the alternative development practices of Turner and Hamdi, the inherent value of Lefebvre's spatial critique of the fragmentation and social production of space can be re-read. In this context, positive difference can be used to confront the negative structural representations of space through the inaction of positive spontaneity and contestation of everyday spatial practices (Massey, 2005, pp. 65, 90). Massey's successful interrogation of the oppressive nature of spatial fetishism and temporal convening confronts the implications of conceiving of space and time in static dualism throughout structural and post-structural theory.[17]

What does positive multiplicity look and feel like? How can we recognise such spaces and more importantly, how can we empower spatial relations and agency that can engage with multiplicity? We have already introduced a number of examples in this chapter and throughout this book, but we must be clear that there is no perfect formula with which to enact positive multiplicity. There is no offer of a panacea here. Instead there is a re-valuing of the chance or in Massey's terms, 'undecidability' and 'throwntogetherness', of real space – of the positive potential of informal and unfinished participations in space.

We thus might just as easily cite the contemporary (and increasingly co-opted) 'pop-up' shops and architectures (Harper and Pawlett Jackson, 2015; Murphy, 2014), which appear to reflect much of the social enterprise and spirit of much innovative informality of the Global South (Cross and Morales, 2013). Yet these merely reflect a narrow spatial interpretation of multiplicity. We could cite the changing faces of London streets like Brick Lane which are continually praised for their diversity and multicultural richness (Hernández, 2010, p. 111), and we would be getting closer to a positive multiplicity but still lacking a methodology of practicable agency that we so dearly aspire towards. And thus here is perhaps a moment of insight where we could begin to suggest that traditional models and methodologies of spatial (architectural) agency are not currently capable of engaging with the multiplicity of space.

In seeking spatial methodologies of agency that might engage with and empower differential space and spatial relations, our comparison of Lefebvre's and Massey's spatial discourses to development methodologies in this chapter is valuable. We can begin to frame theoretical engagements with global inequality and positive heterogeneity as a means to reflect on our own Westernised spaces (Krishnaswamy, 2002, p. 119). By extension our critical comparisons and re-readings of Turner and Hamdi are inherently also an implicit con-testation of Lefebvrian and other Marxist advocacies of political contestations of space. Our earlier critical comparison of Turner and Lefebvre could have been considered somewhat a historical conflation of chance and serendipity. However, this re-reading of Massey's utilisation of difference as the positive foundation of multiplicity, appropriation, and participation clearly reinforces our wider premise for a comparison and a re-contextualisation of both the problems and the positive potential of space within global development (Massey, 2005, p. 39). Yet this raises further difficult questions concerning the identity politics of difference and how we might begin to reflectively learn from such spaces.

Continuous multiplicity, positive difference, and otherness

Exploring the implications and opportunities of multiplicity further, Massey's analysis leads her to utilise both Henri Bergson's (1910) and Gilles Deleuze's (1991) articulations of relational space in terms of 'discrete' or 'continuous' multiplicities. This contestation of difference reflects a continuation of her earlier feminist work exploring the notion of negative difference in gender politics and is part of a critique of Derrida's dualistic notion of 'othering' as negativity (Massey, 1994, pp. 4, 118). The negativity of Derrida's 'other' as conceived as merely the opposite of an accepted identity is confronted by Massey in her examination of the spatial implications of discrete or continuous multiplicity as a socio-spatial condition.

Massey contests the critical distinction between the discrete and continuous interpretations of multiplicity drawn from Bergson, against the inherent negativity of Derrida's *othering*, and observes that they both employ the same forms of dualism of time and space, masculine and feminine (Massey, 2005, p. 144, 1994, p. 260). Her critique highlights how much post-structuralist theory still relies upon discrete and hence closed multiplicities of choice. The pluralism of post-structural space and identity is all too easily construed as simply many more options of structural categorisation of discrete multiplicity. To be truly positive the structural categorisation must give way to open and continuous multiplicity. By exposing that such simplistic dualisms of space are maintained within Bergson's (and subsequently Deleuze's) post-structural theory, Massey reinforces the explicit need for the positive multiplicity of space to counter these negative and oppositional interpretations of post-structural space and time:

> The argument here is instead to understand space as an open ongoing production. As well as injecting temporality into the spatial this also reinvigorates its aspect of discrete multiplicity; for while the closed system is the foundation for the singular universal, opening that up makes room for a genuine multiplicity of trajectories, and thus potentially of voices. It also posits a positive discrete multiplicity against an imagination of space as the product of negative spacing, through the abjection of the other. [...] On this reading neither time nor space is reducible to the other; they are distinct. They are, however, co-implicated. On the side of space, there is the integral temporality of a dynamic simultaneity. On the side of time, there is the necessary production of change through practices of interrelation.
>
> (Massey, 2005, p. 55)

Yet Massey is explicit in her critique of the negativity of deconstruction and is clear that the utilisation of Derrida's identification of post-structural *difference* is positively linked to the political argument for practical, open, and positive space (2005, p. 51). In contrast to the other as a negativity and a definition of a discrete multiplicity as constituted by division and separation, Massey advocates a continuum and plurality of multiplicity as overlapping, open, and continuous (2005, p. 21). Multiplicity is not merely positions on a sliding scale of choices. It is relational, producing identities that interconnected, fluid, and never closed. This is the positive potential of multiplicity in terms of both identity and space. Thus by contesting this connection between difference and otherness Massey activates space as the medium within which to confront identity and pluralism, ideology, and multiplicity. The potential positivity of difference exists in multiplicity as an expression, acceptance, and advocacy of the configurational openness of space (2005, p. 12).

Thus in contrast to Bergson's and subsequently Deleuze's articulation of discrete multiplicity (1991, pp. 44–45), Massey's alternative advocacy of continuous multiplicity as a positive spatial condition intersects with our re-imagining of space and architecture in the context of global and relational comparisons with participatory development practices. Here Massey's utilisation of the Lefebvrian spatial turn provides an interpretation of a socially relational production of space that does not seek to produce multiplicity. Instead it merely engages with and contests the inherent political implications that exist within the material and relational reality of complex space (Massey, 2005, p. 195). Here difference, multiplicity, and otherness can be seen to intersect within the medium of positive space and identity. It is this articulation of the spatial relations of multiplicity that is revealed in our comparisons to development practice. Thus the wider analysis of this research explores how Turner and Hamdi employ alternative socio-spatial methodologies which positively engage in the politics of space, and thus begin to reflectively critique our own Westernised assumptions of inevitability, freedom, and choice (Mouffe, 2013, p. 29).

In Western space multiplicity is limited to places of multiculturalism as an expression of pluralism defined by the co-option and capitalisation of prescribed identities of other cultures, leading to the potential exploitation of these simplified discrete multiplicities. Yet multiplicity as a positive encounter with global and local relational politics of space must be in its very essence a continuous and open medium for change. Continuous multiplicity is not a spectrum of race, culture, class, or nationality. It is not a scale of otherness by which to be measured. It is internally composed of social relations that are interdependently at the scale of the individual, the community, and the world in its entirety. It is open to the point at which it reveals limitless opportunities for change.

This complex distinction of multiplicity as open, continuous, and produced as a spatial practice of identity is of fundamental importance in our explorations of difference and identity regarding development in later chapters. However, the spatial relations and practices of informality and difference must be understood here as not constructed out of a deconstructed negativity or opposition, but from the open continuum of identity and space that is produced by the true complexity of interactions, agonisms, and intersections that are the reality of relational and material space. What this implies with regard our re-imagining of architecture and space is complex. Thus we must look to maintain a reflective criticism of ourselves and take every opportunity to learn from the informal spaces of the Global South by grounding our anlaysis in critical comparisons of Turner's and Hamdi's spatial practices and methodologies.

Lefebvre and Massey – difference and multiplicity

The critical comparisons in this chapter have posited that Massey's articulation of multiplicity intersects with Lefebvre's interconnected concepts of appropriation and differential space. As such it has provided a theoretical framework and lens through which Lefebvre's discourse on the city and difference can be re-read in the context of contemporary global development inequality. As a consequence of this analysis, both Lefebvre's and Massey's advocacies of positive heterogeneity can be contested within the wider confrontation of post-structural global discourses of identity, authority, and difference explored in this book.

This complex trajectory between Lefebvre's and Massey's positive articulations of difference, multiplicity, and the production of alternative socio-spatial practice was perhaps aligned closest with Andrea Cornwall's notion of 'invited spaces'; in participatory development practice. Yet this does not exhaustively or conclusively explore the global potential of Lefebvre's spatial critique, and neither does it confront the problematic implications of introducing the global inequality of development into Lefebvre's discourse (Merrifield, 2006, p. 118). Whilst Lefebvre explicitly recognises uneven development and the predicament of the developing world as an expression of the hegemony of global capitalism, it remains merely a momentary point of critique rather than being contested as the opportunity for change, spontaneity, and positive agency and practice (Lefebvre, 1991, p. 383). Thus it is only in scarce explorations such as Merrifield (2006, p. 130) and Goonewardena et al. (2008a, 2008b), that notions of post-colonialism, identity, and pluralism begin to be connected to Lefebvre's critique of abstract space.

This utilisation of Massey's multiplicity as a global contextualisation of Lefebvre's difference and spatial appropriation provides a foundation and theoretical lineage for the interdisciplinary comparison with Turner and Hamdi in the coming chapters. These later critical comparisons will seek to articulate and compare the positive heterogeneity of development practice methodologies as a mechanism to reveal the true political potential of differential space and multiplicity in spatial practices and social relations of informal space:

Imagining space as always in process, as never a closed system, resonates with an increasingly vocal insistence within political discourses on the genuine openness of the future. It is an insistence founded in an attempt to escape the inevitability which so frequently characterises the grand narratives related by modernity. The frameworks of Progress, of Development and of Modernisation, and the succession of modes of production elaborated within Marxism, all propose scenarios in which the general directions of history, including the future, are known. [...] Only if the future is open is there any ground for a politics which can make a difference.

(Massey, 2005, p. 11)

The above quotation's articulation of the political implications of difference only being possible 'if the future is open' provides a concise and crucial connection to the plural global trajectories of development. By implication this premise contends that only the open development of continuous and practised multiplicity offers a means to transcend the historical, spatial, and political directions of globalisation or the similar restrictions of negative Marxist thought and social agency. The introduction of Massey's positive heterogeneity, openness, and multiplicity of space stands in contrast to institutional Marxism's structural and negative critique, allowing us to articulate space as a medium for positive possibilities and a critical lens through which to compare and critique development and architecture. Massey's introduction of multiplicity can thus be understood as implicating space with the potential of alternative and multiple futures, stories, and social relations, but is also intimately connected to the human acts of social participation that make such alternatives possible. This is in effect the positive heterogeneity or multiplicity that this book seeks to explore, compare, and contest in the remaining chapters.

In these coming chapters, our comparisons to development practice methodologies build upon the critical intersections observed between Lefebvre's and Massey's discourses. These connections offer opportunities for critical reflection on the practical potential and implications of space and the right to difference and multiplicity. Does the openness and necessity of participatory development provide a medium of space in which the social aspirations for difference and multiplicity of Lefebvre and Massey have actually been realised? And if so, what does this imply as a reflective critique of the assumed exemplary realisations of choice and freedom offered by Western spatial relations and practices? The theoretical framework explored here suggests a valuable connection between the right to difference and the positive potential of continuous multiplicity as ways to interpret and understand the global and relational production of space. These connections further reinforce the comparisons and re-readings of development practices offered by our study as a means to critically re-imagine Westernised architecture and space in the context of difference, multiplicity, and otherness.

Notes

1 A similar observation to Marx's reaction to Hegel's abstracted dialectics and his inversion to generate dialectical materialism, as highlighted previously.

2 Allotments in the UK date from as far back as 1732, though the official oldest site dates from 1809 and is in Geat Somefford in Wiltshire. The need for allotments reflects a reaction to the starvation and impoverishment caused to peasant farmers by the Inclosure Acts that occurred from 1773 to 1914. The loss of subsistence farming land contributed to the urban migration and coincided with the industrial revolution. The land available for personal cultivation by the poor was greatly diminished leading to the production of allotment legislation – the Small Holdings Act 1908, and its subsequent modification in 1922 and 1950. Further research on the relation of allotments to the themes raised in this book is tantalising, but must here remain reserved for future research.

3 Sidenote: Is there a potential connection between the urban seasonality of popular sport fixtures as a replacement for the cyclical hope and rebirth offered by agricultural seasonality?

4 The exploration of a Marxist critique of space as a social process remains a provocative re-interpretation and can be read in parallel with many post-Marxist discourses, such as post-colonial theory, which we will utilise later in chapters' comparisons with Homi Bhabha and Gayatri Spivak.

5 Originally published in German – Engels, 1844 – then translated and published in English in 1847.

6 These observations can be compared to a similar evolution observed in development practice. What began as attempts to solve a housing crisis (i.e. Turner) have had to evolve to become explicitly engaged with the social relations of urban cities (i.e. Hamdi). See further observations of this transition in later chapters.

7 Further reflecting the observations and intersections explored in the comparison of Turner and Lefebvre found in the previous chapter.

8 Intriguingly Khedive Ismail commissioned Tahrir Square as part of a 'Paris on the Nile' development programme.

9 See later chapters for more on this distinction between agonism and antagonism.

10 Note the commodification and reification of a political protest into a piece of cultural capital.

11 From 2001 to 2006 Brian Haw lived in his peace camp in Parliament Square, London to protest against the Iraq war. During this time the legislation concerning permission to protest was enacted, meaning that Haw was unable to leave the site without forfeiting his continued right to protest without permission. He was illegally evicted in 2006, after which time Mark Wallinger recreated the protest site and paraphernalia inside the Tate Britain as an artwork.

12 The inability of contemporary Western spatial practices to contest the city is raised in our critical engagement with the concept of cultural hegemony in later chapters.

13 This articulation is made by way of Edward Soja's seminal expansion of Lefebvre's spatial discourse into the hybridity of Los Angeles. The intersection of Soja and Bhabha will not be discussed here as its isolation in the extreme Westernised space of Los Angeles would require a protracted comparison against informal settlements. It does however remain an intriguing opportunity for further research. See Soja, 1996.

14 Open-ended spatial practice is also a theme we encounter later in our study of Nabeel Hamdi's participatory development practices.

15 This reference here to agonism is a connection to Chantal Mouffe and Ernesto Laclau's discourse which this book will connect with extensively in later chapters.

16 For more details see Remi Hess's 'postface' to the third edition of *La survie du capitalisme*, Anthropos, Paris, 2002, pp. 197–214. Lefebvre, 1976.
17 And this analysis provides the foundation for a spatial critique of post-colonial contestations of identity, values and hegemony that will be explored in later chapters.

References

Bakhtin, M., 1999. *Towards a Philosophy of the Act*, Reprinted Edition. Slavic Series. University of Texas Press, Austin.

Baldwin, J., 2012. Putting Massey's Relational Sense of Place to Practice: Labour and the Constitution of Jolly Beach, Antigua, West Indies. *Geografiska Annaler Series B: Human Geography* 94, 207–221.

Bauman, Z., 2000. *Globalization: The Human Consequences*. Columbia University Press, New York.

Bergson, H., 1910. *Time and Free Will*. George Allen and Unwin, London.

Bhabha, H.K., 2004. *The Location of Culture*. Routledge, London.

Bhabha, H.K., 2006. Cultural Diversity and Cultural Differences, in: Ashcroft, B., Griffiths, G., Tiffin, H. (Eds), *The Post-Colonial Studies Reader*. Routledge, New York, pp. 155–157.

Borden, I., 2001. *Skateboarding, Space and the City: Architecture and the Body*. Berg, Oxford.

Borden, I., Pivaro, A., Rendell, J., Kerr, J. (Eds), 2001. *The Unknown City: Contesting Architecture and Social Space*. MIT Press, Cambridge, MA.

Bourdieu, P., 1991. Physischer, sozialer and angeeigneter physischer Raum, in: Wentz, M. (Ed.), *Stadt-Räume*. Campus, Frankfurt.

Bourdieu, P., 2010. *Distinction: A Social Critique of the Judgement of Taste*. Routledge, London.

Chakrabarty, D., 2007. *Provincializing Europe: Postcolonial Thought and Historical Difference*, New Edition. Princeton University Press, Princeton, NJ.

Cooke, B., Kothari, U. (Eds), 2001. *Participation: The New Tyranny?* Zed Books, London.

Cornwall, A., 2002. *Making Spaces, Changing Places: Situating Participation in Development*, Working Paper series 170. IDS, Brighton.

Cornwall, A., 2004. Spaces for Transformation? Reflections on Issues of Power and Difference in Participation in Development, in: Hickey, S., Mohan, G. (Eds), *Participation: From Tyranny to Transformation*. Zed Books, London.

Cross, J.C., Morales, A., 2013. *Street Entrepreneurs: People, Place and Politics in Local and Global Perspective*. Routledge, London.

Crouch, D., Ward, C., 1997. *The Allotment: Its Landscape and Culture*. Five Leaves, Nottingham.

Davis, M., 2007. *Planet of Slums*, Reprinted Edition. Verso, London.

Debord, G., 1994. *The Society of the Spectacle*. Zone Books, New York.

Deleuze, G., 1991. *Bergsonism*. Zone Books, New York.

Derrida, J., 1987. *Positions*, 2nd Edition. Athlone Press, London.

Derrida, J., 1996. Remarks on Deconstruction and Pragmatism, in: Mouffe, C. (Ed.), *Deconstruction and Pragmatism*. Routledge, London.

Derrida, J., 2001. *Writing and Difference*. Routledge, London.

Dikec, M., Dikeç, M., 2012. Space as a Mode of Political Thinking. *Geoforum* 43, 669–676. doi:10.1016/j.geoforum.2012.01.008

Elba, M., 2015. Arrested Development: The 'New Capital' in Egypt. *Journal of Political Inquiry.* Available at http://jpinyu.com/2015/05/04/arrested-development-the-new-capital-in-egypt/ (accessed 9 March 2016).

Elden, S., 2004. *Understanding Henri Lefebvre; Theory and the Possible.* Continuum, London.

Engels, F., 1844. *The Condition of the Working Class in England.* Otto Wigand, Leipzig.

Engels, F., Marx, K., 1975. The Condition of the Working Class in England, in: *The Collected Works of Marx and Engels.* International Publishers, New York.

Evans, A., 2014. On the Origins of Hip Hop: Appropriation and Territorial Control of Urban Space, in: Maudlin, D., Vellinga, M. (Eds), *Consuming Architecture: On the Occupation, Appropriation and Interpretation of Buildings.* Routledge, London.

Fabian, J., 2002. *Time and the Other.* Columbia University Press, New York.

Featherstone, D., Korf, B., 2012. Introduction: Space, Contestation and the Political. *Geoforum* 43, 663–668.

Garrett, B., 2013. *Explore Everything: Place-hacking the City.* Verso, London.

Gilbert, L., Dikec, M., 2008. Right to the City, in: Goodewardena, K., Kipfer, S., Milgrom, R., Schmid, C. (Eds), *Space, Difference, Everyday Life: Reading Henri Lefebvre.* Routledge, New York.

Goonewardena, K., 2008. Marxism and Everyday Life, in: Goodewardena, K., Kipfer, S., Milgrom, R., Schmid, C. (Eds), *Space, Difference, Everyday Life: Reading Henri Lefebvre.* Routledge, New York.

Goonewardena, K., Kipfer, S., Milgrom, R., Schmid, C., 2008a. Globalizing Lefebvre?, in: Goodewardena, K., Kipfer, S., Milgrom, R., Schmid, C. (Eds), *Space, Difference, Everyday Life: Reading Henri Lefebvre.* Routledge, New York.

Goonewardena, K., Kipfer, S., Milgrom, R., Schmid, C. (Eds), 2008b. *Space, Difference, Everyday Life: Reading Henri Lefebvre.* Routledge, New York.

Habermas, J., 1984. *The Theory of Communicative Action, Vol 1: Reason and the Rationalisation of Society.* Beacon Press, Boston, MA.

Hall, S., Massey, D., Rustin, M., 2013. After Neoliberalism: Analysing the Present. *Soundings* 53 (April), 8–22.

Harcourt, W., Brooks, A., Escobar, A., Rocheleau, D., 2013. A Massey Muse, in: Featherstone, D., Painter, J. (Eds), *Spatial Politics: Essays for Doreen Massey.* Wiley-Blackwell, Chichester.

Harper, P., Jackson, P., 2015. The Problem with 'Young Architecture'. *Architectural Review*, 24 February. Available at http://www.architectural-review.com/rethink/the-problem-with-young-architecture/8678914.fullarticle (accessed 9 March 2016).

Harvey, D., 2003. The Right to the City. *International Journal of Urban and Regional Research* 27, 939–941.

Harvey, D., 2012. *Rebel Cities.* Verso, London.

Hernández, F., 2010. *Bhabha for Architects, Thinkers for Architects.* Routledge, London.

Hickey, S., Mohan, G. (Eds), 2004a. *Participation: From Tyranny to Transformation.* Zed Books, London.

Hickey, S., Mohan, G., 2004b. Relocating Participation Within a Radical Politics of Development: Insights from Political Practice, in: *Participation: From Tyranny to Transformation.* Zed Books, London.

Kipfer, S., 2008. How Lefebvre Urbanized Gramsci, in: Goodewardena, K., Kipfer, S., Milgrom, R., Schmid, C. (Eds), *Space, Difference, Everyday Life: Reading Henri Lefebvre.* Routledge, New York.

Knabb, K. (Ed.), 2006. *Situationist International Anthology*, Revised and Expanded Edition. Bureau of Public Secrets, Berkeley, CA.

Koczanowicz, L., 2011. Beyond Dialogue and Antagonism: A Bakhtinian Perspective on the Controversy in Political Theory. *Theory and Society* 40(5), 553–566.

Krishnaswamy, R., 2002. The Criticism of Culture and the Culture of Criticism: At the Intersection of Postcolonialism and Globalization Theory. *Diacritics* 32, 106–126.

Lefebvre, H., 1969. *The Explosion*. Monthly Review Press, New York.

Lefebvre, H., 1970. *Le Manifeste Differentialiste*. Gallimard, Paris.

Lefebvre, H., 1976. *The Survival of Capitalism*. Allison and Busby, London.

Lefebvre, H., 1980. Marxism Exploded. *Review* 4, 19–32.

Lefebvre, H., 1991. *The Production of Space*. Blackwell Publishing, Oxford.

Lefebvre, H., 1996. *Writings on Cities*. Blackwell Publishing, Oxford.

Lefebvre, H., 2003. *The Urban Revolution*. University of Minnesota Press, Minneapolis.

Lefebvre, H., 2008. *La Somme et le Reste*, 4th Edition. Economica, Paris.

Lefebvre, H., 2009. *Henri Lefebvre – State, Space, World: Selected Essays*. University of Minnesota Press, Minneapolis.

Lummis, C.D., 1997. *Radical Democracy*. Cornell University Press, Ithaca, NY.

Lummis, C.D., 2010. Equality, in: Sachs, W., *The Development Dictionary*. Zed Books, London.

Mason, P., 2012. *Why it's Kicking off Everywhere: The New Global Revolutions*. Verso, London.

Massey, D., 1992. Politics and Space/Time. *New Left Review* 196, 65–84.

Massey, D., 1994. *Space, Place and Gender*. Polity Press, Cambridge.

Massey, D., 2004. Geographies of Responsibility. *Geografiska Annaler Series B: Human Geography* 86, 5–18.

Massey, D., 2005. *For Space*. Sage Publications, London.

Massey, D., 2007. *World City*. Polity Press, Cambridge.

Massey, D., 2009. Concepts of Space and Power in Theory and in Political Practice. *Documents d'Anàlisi Geogràfica* 55, 15–26.

McGuirk, J., 2014. *Radical Cities: Across Latin America in Search of a New Architecture*. Verso, London.

McGuirk, J., 2015. Urban Commons have Radical Potential – It's Not Just About Community Gardens. *The Guardian*, 15 June. Available at http://www.theguardian.com/cities/2015/jun/15/urban-common-radical-community-gardens (accessed 9 March 2016).

Merrifield, A., 2006. *Henri Lefebvre – A Critical Introduction*. Routledge, New York.

Miessen, M. (Ed.), 2007. *The Violence of Participation*. Sternberg Press, Berlin.

Miessen, M., 2011. *The Nightmare of Participation (Crossbench Praxis as a Mode of Criticality)*. Sternberg Press, Berlin.

Milgrom, R., 2008. Lucien Kroll: Design, Difference, Everyday Life, in: Goodewardena, K., Kipfer, S., Milgrom, R., Schmid, C. (Eds), *Space, Difference, Everyday Life: Reading Henri Lefebvre*. Routledge, New York.

Mouffe, C., 2013. Space, Hegemony and Radical Critique, in: Featherstone, D., Painter, J. (Eds), *Spatial Politics: Essays for Doreen Massey*. Wiley-Blackwell, Chichester.

Mumford, L., 1961. *The City in History: Its Origins, its Transformations, and its Prospects*. Harcourt, Brace & World, New York.

Murphy, D., 2014. The Pop-up Problem: Does the Obnoxious Phenomenon have Hidden Depths? *The Architectural Review* 16 July.

Nadal-Meslio, S., 2008. Lessons in Surrealism, in: Goodewardena, K., Kipfer, S., Milgrom, R., Schmid, C. (Eds), *Space, Difference, Everyday Life: Reading Henri Lefebvre.* Routledge, New York.

Nancy, J.-L., 1991. *The Inoperative Community.* University of Minnesota Press, Minneapolis.

Neuwirth, R., 2006. *Shadow Cities: A Billion Squatters, A New Urban World,* New Edition. Routledge, New York.

Neuwirth, R., 2012. *Stealth of Nations: The Global Rise of the Informal Economy,* Reprinted Edition. Anchor Books, New York.

Prigge, W., 2008. Reading the Urban Revolution, in: Goodewardena, K., Kipfer, S., Milgrom, R., Schmid, C. (Eds), *Space, Difference, Everyday Life: Reading Henri Lefebvre.* Routledge, New York.

Rahnema, M., 2010. Participation, in: Sachs, W., *The Development Dictionary.* Zed Books, London.

Sachs, W., 2010. *The Development Dictionary. A Guide to Knowledge as Power,* 2nd Edition. Zed Books, London.

Sadler, S., 1999. *The Situationist City,* New Edition. MIT Press, Cambridge, MA.

Saldanha, A., 2013. Power Geometry as Philosophy of Space, in: Featherstone, D., Painter, J. (Eds), *Spatial Politics: Essays for Doreen Massey.* Wiley-Blackwell, Chichester.

Sassen, S., 2001. *The Global City: New York, London, Tokyo,* 2nd Revised Edition. Princeton University Press, Princeton, NJ.

Schmid, C., 2008. Lefebvre's Theory of the Production of Space, in: Goodewardena, K., Kipfer, S., Milgrom, R., Schmid, C. (Eds), *Space, Difference, Everyday Life: Reading Henri Lefebvre.* Routledge, New York.

Schmuely, A., 2008. Totality, Hegemony, Difference: Henri Lefebvre and Raymond Williams, in: Goodewardena, K., Kipfer, S., Milgrom, R., Schmid, C. (Eds), *Space, Difference, Everyday Life: Reading Henri Lefebvre.* Routledge, New York.

Scott, J., 1987. *Weapons of the Weak: Everyday Forms of Peasant Resistance,* New Edition. Yale University Press, New Haven, CT.

Scott, J., 1990. *Domination and the Arts of Resistance: Hidden Transcripts.* Yale University Press, New Haven, CT.

Scott, J., 1998. *Seeing Like a State: How Certain Schemes to Improve the Human Condition Have Failed.* Yale University Press, New Haven, CT.

Shields, R., 1999. *Lefebvre, Love and Struggle: Spatial Dialectics.* Routledge, London.

Soja, E.W., 1996. *Thirdspace: Journeys to Los Angeles and Other Real-and-imagined Places.* Wiley-Blackwell, Oxford.

Spivak, G.C., 1984. Marx after Derrida, in: Cain, W. (Ed.), *Spivak, Philosophical Approaches to Literature: New Essays on Nineteenth and Twentieth Century Texts.* Bucknell University Press, Cranbury, NJ.

Spivak, G.C., 1985. Scattered Speculations on the Question of Value. *Diacritics* 15(4), Marx after Derrida (Winter), 73–95.

Spivak, G.C., 1998. Can the Subaltern Speak?, in: Grossberg, L., Nelson, C. (Eds), *Marxism and the Interpretation of Culture.* Macmillan Education, Basingstoke.

Spivak, G.C., 1999. *A Critique of Postcolonial Reason: Toward a History of the Vanishing Present.* Harvard University Press, Cambridge, MA.

Stanek, L., 2008. Space as Concrete Abstraction, in: Goodewardena, K., Kipfer, S., Milgrom, R., Schmid, C. (Eds), *Space, Difference, Everyday Life: Reading Henri Lefebvre.* Routledge, New York.

3 Geometries of power, hegemony, and small changes

In the last chapter we explored how Massey's contemporary and globalised contextualisation of space offers a new critical lens with which to consider socio-spatial concepts of 'the right to difference', participation and multiplicity, and specifically Lefebvre's spatial appropriation of Marxist theory. This alternative and globalised appropriation of Lefebvrian discourse and analysis offers a foundation from which Massey's critical concepts such as 'geometries of power' and 'cultural hegemony' can now begin to be perceived and contextualised against Hamdi's practices in the Global South. In this chapter we will observe how Massey's analysis and positive advocacy of relational and interdependent global space can thus be critically compared and re-contextualised against specific aspects and methodologies found in Hamdi's development practices and discourse.

By returning to a similar interdisciplinary comparison to that of chapter one's between Turner and Lefebvre, our analysis of the development practice of Hamdi's and Massey's spatial advocacies will highlight critical intersections that are revealed by the concepts of geometries of power, disruption and hegemony, and socio-spatial practices of small change. These comparisons are based upon a close critical reading of the methodological practices and observations of Hamdi as exemplars and concrete realisations of Massey's advocacy of space as a positive and open socio-political development and multiplicity.

In connection with the wider premise of our research, in this chapter we will critically question whether the necessity and scarcity of informal space is inherently valuable when seeking positive and open-ended articulations of socio-spatial relations and alternative development practices that might contest and confront the inevitability of Westernised space. We shall explore how the informality of the Global South is intrinsically connected to Hamdi's articulation and valuing of simple spatial methodologies that generate social change through the contestation of complex spatial relations. Thus, whilst the opportunity for change that Lefebvre observed as held within the social relations of space and difference are revealed by Hamdi's methodologies, it is in the plurality and openness of his approach that Hamdi's practices intersect more directly with Massey's relational multiplicity of space.

Examples drawn from Hamdi's work demonstrate that economic and social disruptions to the space of development can be used to contest and confront the political and cultural hegemony that pervade the Global South (Chibber, 2013, p. 119). Here Massey's premise that the relations, contentions, and conditions of space are relationally produced is confronted by the spatial practices of agonism and disruption (Laclau and Mouffe, 2001, p. 190), and the rich polyphony of possibilities offered by informal spaces and development. This critical comparison of abstract theory to grass-roots practice is further strengthened by the introduction of examples from Westernised space and architecture. These practices and spatial engagements are appropriated here as realisations of the positive potential for participatory development methodologies within the economic and social context of capitalism and the Global North.

Why is this discussion important for our wider attempts to re-imagine space and architecture? How do questions of difference, multiplicity, and otherness relate to Massey's notions of power-geometry, or Hamdi's practices of small change? By comparing such interdisciplinary protagonists the critical analysis and interdisciplinary comparisons in this chapter provide previously unheard of connections between spatial theory discourse that targets positive political agency and practical real-world methodologies and achievements drawn from the Global South. The juxtaposition of these ideas, concepts, and strands provides an invaluable chance to see the true potential of space and architecture from the perspective of 'others'. These comparisons provide the opportunity to re-imagine spatial agency against the disruptive and innovative participatory practices of grass-roots development, and our underlying agenda to confront the positive potential of difference, and multiplicity, and otherness of space.

This chapter is also structured to provide theoretical introductions to the various concepts that define Massey's spatial advocacies. These concepts are primarily drawn from her seminal text *For Space* (2005), but also explore the academic papers and articles that accompanied her writing at this time, and the other key discourses of seminal authors that Massey critiques and re-appropriates extensively in her work. Thus we will utilise Massey's key concepts of multiplicity, relationality, and specificity and explore their connection with space, place, and time.

Building on these positive spatial concepts we will explore Massey's 'geometries of power' and the spatial articulations of Gramscian cultural hegemony, utilising Hamdi's practices of small change as concrete examples of development as a positive disruptive force. In doing so this chapter will analyse the implications of Hamdi's open-ended and speculative methodologies as a critical cypher of concepts that Massey explores in the wake of defining the positivity of agonistic practices: namely, the undecidability, specificity, and relational interdependence of space.

Whilst these concepts and relationships are complex, the ability to critically compare them to the overtly simple and humble practices of Hamdi is vital to both ground and expand the impact of Massey's positive spatial advocacy. In addition to examples of Hamdi's practices, this chapter will also introduce various examples drawn from the small number of alternative architectural and spatial practices that have been explored in contemporary Westernised architecture and space. This combination of recognisable examples and spatial methodologies will provide an interconnected mesh of relational contexts with which to ground the otherwise somewhat intangible concepts of spatial multiplicity, agonism, and undecidability.

As with our comparison of Turner and Lefebvre, the value of discussing Massey in the pursuit of alternative models of architecture and development is highlighted by the juxtapositions of theory and practice, and the Global North and South. In exploring Hamdi's almost counter-intuitively simple and humble spatial practices this chapter begins to reveal contemporary methodologies that can change the way we – architects, bureaucrats, economists, the public, and so on – produce space. Considered reflexively against conventional architectural practices the contrasts and contradictions of these practices begin to frame a critique of the ideological inevitability of Westernised space.

Hegemony and space

In order to further explore and critically compare and re-contextualise Hamdi's practices of 'small change', and our broader attempts to unravel a critical discourse on the inevitability of development, Massey's utilisation of spatial hegemony must first be contextualised by pursuing its broader political and cultural foundations.

This process begins with a brief explication of Marx's descriptions of hegemony, which can be reasonably and succinctly summarised as the systemic oppression of the working class by a ruling elite through ideology and super-structure, and the cultural institutions, power structures, and rituals of the state (Laclau and Mouffe, 2001, p. 85; Marx, 1977, p. 164). Later, Antonio Gramsci would famously develop this conceptual articulation of the inter-dependence of political frameworks and socio-economic inequality in his examination of 'cultural hegemony' and its implications as an explicitly cultural sphere of intellectual and moral leadership concurrent with politics (Jones, 2006, pp. 49–52).

Gramsci critiqued hegemony as cultural practices of identity, institutional representation, and fundamentally as the suppression of alterity and otherness (Krishnaswamy, 2002, p. 115). However, he was also very careful to articulate hegemony as not defining an unchangeable inevitability but merely reflecting the implications of a dominant cultural power. This analysis suggests a com-plementary Marxist interrogation of space to Lefebvre's articulation of the inherently positive potential of the reproduction of the social relations of production as noted in the last chapter (Lefebvre, 1976). Crucially Gramsci notes that:

> it is precisely the porosity of a hegemonic bloc to the demands of others which provides a cause for optimism. A ruling power that asks for

consent and yet which cannot give voice to the aspirations of those in whose name it rules will not survive indefinitely.

(Jones, 2006, p. 47)

Once again reflecting Lefebvre's conception of differential space, this positive conception of the inherent 'porosity' of hegemony also implies that the process of hegemonic development must be continuous, unfixed, and open. In contrast to the inherently false appearance of cohesion that sustains such hegemonic relations, space is rich with identities and communities that represent alternative and subaltern social relations. Gramsci posits that over time such identities have the potential to pass from isolation and exclusion to become protagonists, and eventually as potentially effective counter-movements to the cultural institutions and political ideology (1971, p. 170).

It is this positive potential of hegemonic porosity, articulated through the voices, spaces, and politics of otherness and alterity, that provided the foundation for key aspects of Laclau and Mouffe's now seminal text *Hegemony and Socialist Strategy* (2001). In their aspiration to explore the positive potential of a socialist hegemonic strategy, Laclau and Mouffe build upon Gramsci's critique of the historical sedimentation of Marxism and socialist political theory, which they suggest has become suffocated by a layered historical contingency with capitalism. In contrast to this they propose a challenge to the 'increasing gap between the realities of contemporary capitalism and what Marxism could legitimately subsume under its own categories' (2001, p. viii).

In response to this 'gap', Laclau and Mouffe advocate the necessity of a political reactivation 'to show the original contingency to the synthesis that the Marxian categories attempted to establish' (2001, p. xvii). In other words, their project is an explicit attempt to return to the original political reaction of Marxism against the inherent crises of capitalism. In this task they appropriate Gramsci's departures from institutional Marxism, offering a renewed intellectual arsenal of concepts, and specifically *cultural hegemony*, from which to pursue the potential of an alternative socialist and counter-hegemonic strategy (2001, p. ix).

Gramsci's notion of hegemony as being a political and cultural response to the inherent fracture, rupture, and dislocation of logical capitalist cohesion articulates neoliberal models of development as a manipulation of material and social relations within the progression of historical necessity. This gap or fracture produced by capitalist economics and political culture is both caused by and subsequently proliferates within capitalist space:

> The concept of hegemony did not emerge to define a new type of relation in its specific identity, but to fill a hiatus that had opened in the chain of historical necessity. 'Hegemony' will allude to an absent totality, and to the diverse attempts at recomposition and re-articulation which, in overcoming this original absence, made it possible for struggles to be given a meaning and for historical forces to be endowed with full positivity. The

contexts in which the concept appear will be those of a fault (in the geo-logical sense), of a fissure that had to be filled up, of a contingency that had to be overcome. 'Hegemony' will not be the majestic unfolding of identity but the response to a crisis.

(Laclau and Mouffe, 2001, p. 7)

Thus, capitalist hegemony can be critiqued as the reaction to an imbalance of social and economic relations. It is a cultural and social condition and an articulation of the logical cohesion of capitalist space as inevitability. However as Gramsci, Laclau and Mouffe, and Massey have each sought to articulate, positive alternatives to this inevitability can be found in the subaltern identities, difference, and otherness that exist within these gaps and fractures of capitalist cohesion (Massey, 1994, p. 70; Willis, 2013, p. 132).

Gramsci articulates the hidden yet inherent political and cultural production of hegemony as itself a continuous (dialectic) process. This allows Laclau and Mouffe a provocative theoretical framework from which to contest the limits of hegemony. They propose that if the cultural implications of hegemony lie in the relationships that exist between constructed unequal power-relations and the project of capitalism to produce them, then the opportunity for positive counter-spaces must also originate from such socio-spatial relations of false cohesion (Barnett, 2004, p. 515; Laclau and Mouffe, 2001, pp. 51–52).

In contrast, the agonistic politics posited by Laclau and Mouffe offer a confrontation to hegemonic space, and provide Massey with a positive articula-tion of counter-hegemonic space as interdependent with issues of identity, otherness, and disruption (Rustin, 2013, p. 59). Once again, this analysis of the intersection of hegemony and otherness reinforces the potential of our com-parison of Hamdi's practices of intervention within just such peripheral contexts of alterity in the informal Global South.

Gramsci's 'porosity of hegemony' suggests that alternative constructions of relations and power-geometries could – and indeed for Laclau and Mouffe should – be equally used in a positive way (Jones, 2006, p. 130; Laclau and Mouffe, 2001, p. 183). This point is crucial to our positive comparisons with Hamdi's practices of disruption and small change. If the social and spatial hegemony encountered in development and in all spatial practices is under-stood as a constructed imbalance, then alternative and disruptive practices could be articulated to produce new counter-hegemonic political spaces.

Such spaces are not a panacea. Be they informal settlements in the Global South like the much publicised Torre David in Caracas (Baan et al., 2012),[1] or alternative socio-cultural movements in the Middle East such as the global Occupy movement, Arab Spring political uprisings, or even cultural phenomena like the Slow Food movement (Bower, 2016), none offer a solution for space. But in contrast to conventionally abstract and isolated Westernised space, they might exist as imperfect articulations of more socially viable geometries of power that are practised and performed in explicitly political, plural, and agonistic forms of space. Thus, they might also help to propose a re-imagining of space as a process of positive, open and self-aware spatial relations and the potential of a more socially articulated cultural hegemony: 'Our approach is grounded in privileging the moment of political articulation, and the central category of political analysis is, in our view, hegemony' (Laclau and Mouffe, 2001, p. x).

Laclau and Mouffe explicitly define the political moment and action that is inherent within hegemony as a positive cultural expression of the political voice of an active society. In critique of this perhaps over-simplistic and theoretical explication of positive hegemony, John Clarke is notable as suggesting an almost utopian evasion of the material reality of such positive conceptions. Clarke notes an ideological (Laclau and Mouffe would use *symbolic* (Smith, 1998, p. 103)) and potentially naïve overlooking of the ruling bloc, and capitalism's inherent ability to reshape the conditions upon which such potential alternative actions need to gain momentum (Clarke, 1991).

Similar contestations of the potential political limitations and implications of positive hegemony are articulated by Stefan Kipfer (2008, pp. 206–207), noting hegemony must be understood as forever entwined in a continuous dialectical process. Thus, the proposal that Hamdi's methodologies are comparable exemplars of positive counter-hegemonic spaces is also implicated in the observation that space must always remain dependent upon a continuous process of political democracy and the plural logic of difference and discursive identity (Kipfer, 2008, p. 203). In returning to Massey's spatial contextualisation of the hegemonic process we can see this endless debate as in fact integral to the articulation of relational yet specific equalities of space:

> In order to respond to specificity, however, one needs (ever provisional) agreement about aims, and that requires global fora of a very different nature. [...] The objection to such a suggestion would undoubtedly be that it would lead to endless debate and disagreement. And it undoubtedly would. But endless debate and disagreement are precisely the stuff of politics and democracy.
>
> (Massey, 2005, p. 103)

Building upon the implications of the endless debate implied by open, positive, and politically agonistic space, our notional comparison of positive hegemony and Hamdi's development practice is thus grounded by both Ernesto Laclau's reflections on the positive potential of counter-hegemonic processes (1990, p. 72), and Massey's subsequent critical relational and spatial theoretical field (2005, p. 25). In Massey we find a renewed spatial aspiration and positive re-articulation of the question of hegemony within the interdependence of space and development, and the openness of relations are constantly reproduced. Relational space is contiguous with the continuously shifting power-geometries of socio-political relationships. Thus, the potential of positive hegemony and relational space can be perceived in the methodologies of performance and practice found in both Hamdi's examples of participatory development, and unconventional examples drawn from Westernised space (Massey, 2005, p. 85).

In contrast to today's hegemonic story of globalisation and its temporal convening towards universal and apparently inevitable ideologies of modernity, for Massey global space is about contemporaneity, chance, and positive multiplicity. In contrast to inevitability, space is about openness and must be practised in ways that reveal, contest, and confront the existing relations, fractures, and practices of capitalist hegemonic space. This analysis of positive hegemony and spatial practice allows this chapter to begin to critically compare Hamdi's alternative development methods of small change as concrete exemplars of positive counter-hegemony. Such participatory practices in informal communities across the Global South articulate the potential of counter-hegemonic spaces as contestations of the inevitability of capitalism.

Geometries of power, inevitability, and small change

Having interrogated the concept and implications of hegemony from Marx to Gramsci and on to Laclau and Mouffe, it is now important to examine how Massey develops these ideas further in her work, and how they can be re-appropriated in comparison to the work of Hamdi and others.

Throughout the 1990s Massey produced a series of papers in which she described the concept and implications of 'geometries of power' (1991, 1993, 1999). This term is a foundational concept that interconnects her earlier works on the spatial divisions of labour in the UK (1984), her exploration of space and gender in the 1990s (1994), and her recent discourse on globalised spaces, relationships, and the cities (2007, 2005). Power geometry is a conceptualisation of how the interconnection of power and the globalisation of space effects people and places differently because of gradients of inequality. These effects both depend upon and reflexively generate the inequalities of socio-spatial relationships that afflict global and local spaces alike.

In seeking to unravel and explain the principles of geometries of power, Massey describes how structuralist space is conceived of as a flat, uniform, almost barren surface (1994, p. 4). Devoid of contours, layers, or spaces of agonistic confrontation, this surface conception of space allows structuralist discourse to overlook the inequality and complications of real space. In structural theory, space is merely the medium in which time unfolds: merely the construct in which moments of existence that unfold through time are captured (Massey, 2005, p. 24).

In contrast, Massey's is a conception of space with contours and confrontations. In advocating the complexity of space as interdependent with time in defining place, she describes how space reveals the power struggles and hegemonies that are intrinsic within contemporary society (2005, p. 153). In essence, the inequality of space, global or local, economic, political, or social, are all experienced and thus defined by open and positive space that is interconnected and interdependent with time.

For example, Massey describes the power geometry that is expressed in the Canary Wharf Docklands development in London, and specifically the local inhabitants who were either relocated or now live within the shadow of physical manifestations of power geometry (2007, pp. 7, 29, 2005, pp. 166–169). The neoliberal freedom of development afforded to the London docklands projects is a projection of the power and hegemonic control affecting contemporary Western space.[2] These geometries of power and their interdependent socio-cultural hegemonies reflect the same dichotomies that are observed within the binary constructions of identities: us and them, centre and periphery, Global North and South, the haves and the have nots (Harvey, 2012, pp. 155–157; Massey, 2007, p. 9).

Concurrently, Nabeel Hamdi's seminal texts *Small Change* (2004) and *The Placemaker's Guide to Building Community* (2010) make repeated and explicit reference to the implications of his participatory development practice on the political stability and social fabric of place. These references highlight the social and political implications of such models of development interactions in 'other' people's space (Hamdi, 2004, p. 56; Kaplan, 1996, p. 107), and reflect an explicit recognition of global development intervention in terms of anthropological authority, post-colonial identity, and textual values. These issues are discussed further in chapters four and five, however this chapter first seeks to explore and contextualise Hamdi's methodologies of participatory development in comparison with discourses concerning hegemony, power-geometries, and the practices of agonistic socio-spatial disruption.

Hamdi's development practice methodologies are inherently based upon various techniques he identifies under the concept of 'small change'. These practices of participation, empowerment, and facilitation are focused upon targeted, efficient, and simple practical actions to confront and engage people in the production of social space and the practice of social relations (Hamdi, 2004, p. xx). His discourse and methodologies have been widely applauded and advocated as offering concrete and practical potential for profound spatial and ideological change by placemaking and community building (Burnell, 2012, pp. 135–137; Fraser, 2012, pp. 63, 65).

This chapter posits that these practices also inherently question and contest the cultural assumptions of what change and development look and feel like. They challenge both the institutional assumptions of development and also the expectations of local people, prompting a positive and alternative contestation of how to define and facilitate positive social change (Capra, 2002, p. 102). Hamdi's participatory practices thus offer an alternative vision of change and growth and question the inevitability of development. In contrast to the assumptions of development as a process of homogeneity and inevitability, and in spite of the complexity of the existing social context, Hamdi's practices actively seek to reveal and activate a contestation of existing hegemonic social relations:

Practice disturbs. It can and does promote one set of truths, belief systems, values, norms, rituals, powers and gender relations in place of others. It can impose habits, routines and technologies that may lead to new and unfamiliar ways of thinking, doing and organising, locally, nationally and even globally. It may do this intentionally because the existing structures have become malignant, or because they could work more effectively if they were to change, or because there is no order – no sophistication where it is needed. It may also do so in the interests of one power elite over another to induce internationally a new global order. In all these respects, practice – that artful skill of making things happen; of making informed choices and creating opportunities for change in a messy and unequal world – is a form of activism and demands entrepreneurship.

(Hamdi, 2004, p. xix)

For Hamdi it is clear that being engaged in development practice and actively interested in the spaces of others must inevitability disturb their space. Within this analysis Hamdi provides a tacit recognition that his development practice is engaged with the articulation and manipulation of structures, be it intervening against 'malignant relationships' or introducing 'order and sophistication'. This interpretation of development as inherently an intervention within the hegemony, inevitability, and assumed trajectory of development suggests the concurrent realisation that such contestations inevitably generate both 'winners' and 'losers' (Capra, 2002, p. 102; Hamdi, 2004, pp. 73, 99).

Thus, in the framework of our wider critical analysis, Hamdi's conscious, practical, and agonistic engagement with the disruption of social relations is interpreted as a contestation of social hegemony and spatial inevitability. By engaging in the alternative and differential practices of informal necessity Hamdi's practices of small change and participation offer opportunities for positive disruption of expectations and prejudices, authorities and assumptions, vulnerability and aspirations. Yet in the context of Cornwall's warnings of the political power of invited spaces of participation observed in the previous chapter (2004, p. 80), it is crucial to explore the more complex political and cultural implications of the disruption and the social implications of participation in closer detail (Burgess et al., 1997; Latouche, 2010; Sachs, 2010).

Our interpretation of Hamdi's practices and methodologies as inherently agonistic and disruptive affords a comparison with Massey's critique of the spatial interdependence with power, hegemony, and inevitability. The self-awareness of Hamdi in his articulation of development as a disruptive practice implicates his work with a critical understanding of the inevitable consequence of contesting conditions, relations, and existing issues that define the power geometry and hegemony of his working contexts throughout the Global South. In order to invoke the kinds of social and political change that Hamdi is lauded for achieving, this critical recognition of the socio-political complexity of development suggests that potentially all development practice must be interpreted as an inevitable disruption to the cultural, political, and economic status quo.

This comparative re-reading of Hamdi in the context of Massey's multiplicity of space is built upon explicit recognition of the widely demonstrated implications of capitalism as producing unequal relations not merely of economics, but of geometries of power (be they economic, social, or political) for the benefit of some and to the detriment of others (Harvey, 2012, p. 162, 2010, pp. 75, 78–79; Massey, 1999, p. 33; Saldanha, 2013, p. 48). Once this notion of development as being intrinsically interdependent with both local and global geometries of power is accepted, the remaining question relates to which participants or beneficiaries in such practices are going to benefit from such interventions. The true social and political implications of development practice are revealed in either the positive or negative political articulations and confrontations of such disruptions. In this context Hamdi cites Joke Schrijvers: 'This struggle for world hegemony was and continues to be at the core of what is lovingly referred to as development cooperation', a process in which the poor (and their governments) had to be willing to cooperate if they were to reap the benefits of globalisation and the good life:

> The results: most who participated became co-opted into systems of production and trade, agreed internationally and reflected in such policies as structural adjustment. In practice, the highest toll (of such structural adjustment programmes) fell on the poorest social group, not on governments or other elites. Women, responsible for day-to-day survival and for the children, shouldered the greatest burden.
>
> (Schrijvers, 1993, p. 11)

The notion that within development there must inherently be winners and losers of such interventions is inevitably an oversimplification of the complex situation in the balance of development and global inequality. Yet in contrast to traditional Westernised development interventions of formal and abstract paternalism, by contesting assumptions of development our emerging critical comparison observes that Hamdi in fact advocates an innovative and poststructural approach to development by engaging and valuing the spaces and values of others:

> We had learnt that placemaking could mediate the interests and values [...] of different kinds of community. Engaging these partners in participation would be to 'dance with conflict' literally and metaphorically to acknowledge their roles as agents of change.
>
> (Hamdi, 2010a, p. 33)

In this context it is crucial to observe the intersection of Hamdi's development as disruption and Massey's articulation of the inevitability of development as interdependent with an increasingly homogenous and hegemonic global presumption of space (2005, p. 11, 2004, p. 13). To ground this comparison it is necessary to understand that Massey's contestation of the passivity of such spaces of development is defined by the interdependence of cultural and political relations and the assumption that capitalist models of growth, economics, and society were an inevitable and universal answer:

> Moreover, within the history of modernity there was also developed a particular hegemonic understanding of the nature of space itself, and of the relation between space and society. One characteristic of this was an assumption of isomorphism between space/place on one hand and society/culture on the other. [...] It was a way of imagining space – a geographical imagination – integral to what was to become a project for organising global space. [...] It is a response which takes on trust a story about space which in its period of hegemony not only legitimised a whole imperialist era of territorialisation but which also, in a much deeper sense, was a way of taming the spatial.
>
> (Massey, 2005, p. 64)

This critical observation of the taming of space and global development implies an interdependent link between the hegemony of capitalist social relations and an inevitable cultural passivity and acceptance of what change looks and feels like (Massey, 2008, pp. 496–497; Sachs, 2010). In reading Massey's alternative and critical discourse on space we can contend that modernity as a project conflated the representation and abstraction of space into intersecting trajectories, geographies, and geometries of power (Massey, 2005, p. 63). Reflecting much of Lefebvre's criticism of the fragmentation of abstract space (1996, p. 188), Massey observes and articulates modern space and its concurrent neoliberal political and economic context as a system of formal and abstract structures of assumption, fixity, and inevitability. It is this socio-cultural and political assumption of inevitability that articulates and enforces hegemonic ideologies through space. This critical interrogation of structuralism implies modern space as structurally refusing to acknowledge the fractures, instabilities, and multiplicity of space and culture.

However the full spatial implications of this structural project of modernity have perhaps only begun to be recognised in the wake of post-colonial contestations of identity as complex, overlapping, and practised (Sachs, 2010; Sennett, 2003). Post-colonialism's implied contestation of the cohesion and hegemony of modern space can in some ways be recognised as a reactionary mechanism to deal with the creativity, difference, and confrontation of otherness (Massey, 2005, p. 65). In contemporary post-structural discourse, identity and space are newly recognised as interdependent within complex social and spatial relations that overlap, intersect, and combine to produce a rich multiplicity of space.

In the analysis and comparisons observed here, Massey provides an invaluable point of theoretical critique between the global space of development and the political and cultural hegemony of capitalist policies and practices. These practices advocate and even celebrate the necessity of cohesion and stability within the identity of global development space and concurrently the eventual target of Westernised space itself (Bauman, 2000, pp. 59–60). Thus, whilst Massey's positive multiplicity of space is only able to confront and contest the institutional assumptions of formal abstract Westernised space and structuralist concepts in a theoretical way, Hamdi's practices of disruption suggest the positive potential of a model of alternative, grass-roots, and participatory democratic development that revels in the confrontation of such hegemony (Hall et al., 2013, pp. 8–22).

Small change

> 'small' because that's usually how big things start; 'change', because that's what development is essentially about; and 'small change', because this can be done without the millions typically spent on programmes and projects.
>
> (Hamdi, 2004, p. xxiii)

Hamdi's disarmingly modest title and description of *Small Change* is itself an emblematic articulation of his alternative practice. The humility of this phrase conceals a socio-spatial complexity of methodology that this chapter advocates as a previously unrecognised articulation and practical confrontation of cultural hegemony in development.

As we have observed, Hamdi's engagement with the positive potential of spatial practices utilises a targeted agency of disruption. In contrast to classical Westernised notions of spatial appropriation and transgression against authority through class upheaval, protest, or antagonism, Hamdi's disruptions are distinctively small, humble, and appear at first glance to be largely passive. They are agonistic instead of antagonistic, inciting a social questioning of space and relations as an on-going process, rather than a moment of temporary abstract violence and anger. Hamdi's small change practices act as catalysts that potentially reveal inequality and prompt change through enacting or supporting alternative social relations and spatial practices (Hamdi, 2004, p. xxiii):

> Small Change thinking predominately focuses on placemaking and the transformative way that place-based interventions can generate opportunities for social and economic development. Small Change starts with practice, drawing on local innovation, creativity and entrepreneurship to catalyse change. Through participatory planning, a process is facilitated by which community collectives make important project decisions, including identifying key problems and opportunities, establishing goals and priorities and defining project resources and constraints. Decisions made during this facilitated process direct or are incorporated into traditional placemaking, including architectural design and urban planning. This way of working challenges many professional working practices by raising questions about the amount of formal structure required to successfully deliver community improvement programmes before the structure itself restricts

progress, becomes self-serving and inhibits personal freedom. Small Change thinking also extends beyond place-based interventions to address issues including community-led DRR along with community health and wellbeing initiatives.

(Burnell, 2012, p. 139)

Small change is an articulation of intelligent and creative problem solving that uses design to support, facilitate, and generate open-ended and (crucially) self-driven reliance and social sustainability (Hamdi, 2004, p. 102). Based upon direct observations and experiences, Hamdi's development methodologies are inherently built upon distributed networks of grass-roots social change and upon 'the collective wisdom of the streets' (2004, p. xviii). Alternative development practices such as financial micro-loans in rural areas, or grass-roots women's banking initiatives have not conventionally been considered in comparison with architecture or Westernised space (Esteva and Prakash, 1998, pp. 60, 192–193; Harcourt et al., 2013, p. 160). Yet the socio-spatial challenge that such spatial agency and facilitation poses can be seen to resonate with contemporary alternative spatial practitioners and agents of social change in the Global North.

In terms of such practices in the Global North we might cite the works of the AOC, Assemble studio, Architecture 00, and UTT amongst others, and the passionate recent discourses from Dan Hill (2015) and Marcus Westbury (2015). And analysis of such examples from the Global North benefit from repeated critical comparison to the practices and observations of Hamdi, where simple small change practices confront, contest, and contrast the political, social, and economic interventions of prescriptive top-down models of development (2004, pp. 16–17).

Considering the principles of small change outlined above by Hamdi, Lefebvre's and Massey's Marxist critiques of neoliberal and capitalist space might begin to be re-read in the context of the material conditions of necessity and inequality found in the Global South. These social and material conditions are compounded and exacerbated by conventional abstract assumptions of what development means, and what it looks and feels like. Formal projections of housing and planning reflect an economic, social, and structural implausibility of capitalist development as producing anything other than further inequality (Lummis, 2010, p. 44; Massey, 2005, p. 87). In contrast to practices of small change, conventional development models based upon homogenous space and social hegemony are intrinsically linked to socio-political assumptions of development that only produce further inequality and at an extremely high cost to those who can least afford it (Turner, 1976, pp. 62–66, 127).

In contrast to the homogenous tendencies of contemporary neoliberal globalised development, Hamdi observes a convergence between the principles and methodologies small change, and theoretical discourse concerning emergence and complexity. Using observations and concrete examples he contextualises the importance of producing space and building dense interconnected networks using simple elements (2004, p. xviii). Such networks of inter-changeable materials and relations create the potential for sophisticated, diverse, and socially sustainable economic behaviour to trickle up, rather than be forced down (2004, p. 73). This sentiment and methodology echoes Turner's advocacy of autonomy and heteronomy (1976, pp. 17–19), and the necessity that in order to generate positive change it is more effective and valuable to start small and start where it really counts – in the specificity and material reality of complex contexts and practice (Boano, 2013).

The challenge and opportunity of such examples is in interpreting them in contrast with traditional models of physical intervention. Small change methodologies challenge conventional assumptions of the necessity of devel-opment projects to deliver immediate social impact versus Hamdi's models of intergenerational change and sustainable livelihoods. This is particularly challenging when considering these methodologies in relation to a Western context.

Notwithstanding the increasingly audible acknowledgements of architecture's fated co-option as a part of capitalism (de Graaf, 2015), the theoretical and practical comparisons drawn in this book resonate with an aspiration to re-imagine space as open and positive, and to re-articulate architecture as a verb and a socio-spatial catalyst and agency for change. Here once again this book would highlight the alternative architectural and social agency of actors such as Alistair Parvin and Architecture 00's call for architecture to serve the 'other 99%' of people and spaces that it has traditionally been disconnected from in the contemporary profession (2013).

In order to contextualise the power and value of his small change practices, Hamdi articulates various exemplars. At first these examples sound intractably remote, small, and abstracted from the grand visions of positive political change aspired to in this book. Such examples include the facilitation and support of rubbish pickers, bus stops, recycling sorters, and water-tap attendants, with each example offering the potential for strategic change through social entre-preneurship, community networks, and civic participation in grass-roots eco-nomic and political processes (Hamdi, 2004, pp. 79–82). For example, Hamdi articulates the positive opportunities possible by simply supporting a local composting bin programme that had the potential to be scaled up beyond its initial conception. Rather than starting with a city-wide initiative to compost and recycle, Hamdi's small change practice encouraged a small catalyst project which proved the concept worked and thus could eventually become engaged not only with local authority waste collection policies, but also with education, food, health, and sanitation programmes (2004, pp. 34–35).

In contrast to formal, centralised, and institutionalised development models, small change practices reflect an alternative perception of develop-ment. The expectations of what development is, what it looks like and means, are confronted by Hamdi's acute attention to small, practicable, and efficient change. Such targeted and strategic change challenges people, space, and communities to generate far richer and more densely interconnected social relations that are not reliant upon continuous external aid and support (Hamdi, 2010a, p. 22).

Hamdi's advocacy of practices of small change articulates opportunities to break down assumptions and contest the inevitability of development. This suggests a level of spatial interrogation and impact that is palpably more politically positive and provocative than contemporary institutionalised development in the Global South. It is also suggestive in relation to our wider questioning of the spatial practices and architecture of the Global North which continue to be largely restricted and constrained by hegemonic social and spatial relations. Architecture in the Global North can undoubtedly learn from the positive potential of small change and spatial practices that actively disrupt and contest the assumptions and inevitability of the social relations that define and contain existing hegemonies. Crucially they do not rely upon more money or social upheaval, but revel in humility, efficiency, and intelligent social practice.

Useful examples of such small change methodologies can be identified in the Global North, though they only emerge from very particular socio-economic models of architectural practice. Contemporary examples could include the work of Studio Weave in London, and specifically their competition work such as the 'Watering Poles' for the AJ Kiosk challenge in London, and their public realm projects at Hornchurch, Romford, Croydon, and Hackney.

In citing such projects it is important to retain a sense of objective criticism regarding the work of such practices. Whilst their work does exhibit many positive tenets of small change methodologies, there is doubt as to the economic and social sustainability of such practices, and an awareness that their model of practice is not a panacea and cannot be proliferated without systemic changes to Westernised space, politics, and society.

Small change practices are inherently agonistic in their challenge to the assumptions of space and socio-cultural hegemony that were revealed in Massey's theoretical critique of structuralist space. By critically comparing Hamdi's work with Massey's discourse, his methodologies can begin to be seen not only as simple, practical techniques for engaging with complex socio-political contexts, but more profoundly as perhaps a shift from a process of hegemonic and homogenous development to a process of open change.

Yet the challenge of these methodologies to Westernised contexts remains unquestionably daunting, specifically because of the wider social and political questions that they imply. Why do interventions like that of Studio Weave remain isolated, ephemeral, and largely unable to generate strategic change? Are their methodologies so different to that of Hamdi? Or is Westernised space imbued with far more restrictive cultural relationships of hegemony and power geometry that cannot be transcended? To re-imagine spatial agency through the lens of small change thus reveals both the positive potential of space but also the fundamental challenge that the cultural hegemony of capitalist Westernised space creates.

Positive counter-hegemonic disruption

As explored in our analysis of small change, Hamdi explicitly observes the inherently disruptive nature of practice in terms of the hegemonic inevitability of space (2004, p. 56). However, in response to this inevitability he actively advocates and engages in methodologies of participation that highlight and provoke issues of political agonism (2004, p. 140). Here the political implications of engaging and intervening in informal contexts become an opportunity to react against the constraints of capitalist hegemonic spatial relations and what Massey has described as the 'temporal convening' of space and development (2005, p. 65).

In advocating the necessity of practical methodologies that contest political and social assumptions and relations, Hamdi articulates what could be described as a post-modern self-awareness of his practices in contrast to the negative implications of disruptive development as renewed post-colonial intervention (2004, p. 63; see also: Hickey and Mohan, 2004, p. 61; Kelly, 2004, p. 213). Echoing the global and local relationality of space advocated by Massey (2005, p. 102), these are practices built upon the significance of specific cultural and political participation as a means to empower communities to confront assumptions of their socio-cultural and economic trajectory of development. Thus when describing participatory practices and workshops in informal settlements Hamdi clearly recognises that participation alone is inherently capable of generating or reinforcing relationships and social power geometries:

> Participation [...] often serves to reinforce existing leadership structures; gives dominance to the majority or elite and either way can exclude minorities. It winds up being oppressive to minorities and undermines the sense of belonging.
>
> (Hamdi, 2010a, p. 99)

Such questioning of the problematic positive and negative potential of disruption and intervention within complex cultural contexts continues to underpin Hamdi's discourse. He goes on to note how his practices have to be designed:

> to give definition to the term participation from the points of view of some of the principal actors in development, in order to reveal some of the conflicting agendas and also the complementarity. Moderating the dominance of one actor's agenda versus another, converging interests and negotiating priorities is one of the key roles of facilitation.
>
> (Hamdi, 2010a, p. 87)

Within such observations it becomes clear that the political, economic, and social complexity of informal communities articulates development as an inevitable engagement with the disruption and contestation of social relations. This once again highlights our underlying premise that such relations can be considered as part of a global hegemony of inevitability as described by Massey. Thus, Hamdi's methodologies can be seen as attempts to balance the conflict that exists between the inherent hegemony of what Fritjof Capra described as the necessary social 'willingness to change' (2002, p. 102), and Massey's advocacy of open development and multiplicity (2005, p. 95).

> Participatory programs, in the early stages of planning, also help identify areas of potential conflict among groups vying for power or competing for resources. They tap the ingenuity of people to discover ways of solving problems that may not be a part of the expert repertoire. They enable [...] the construction of alternative versions of the world, to fashion networks of solidarity, and build people's confidence in their own knowledge and capabilities and with it a sense of entitlement.
>
> (Hamdi, 2010a, p. 93)

Participatory practice inherently recognises disruption as an inevitability of development. Yet crucially Hamdi utilises these practices as opportunities for all actors and agencies to discuss, reveal, and recognise the socio-political and economic relations that might need to be questioned, challenged, and disrupted. Thus, Hamdi's methodological use of disruption intrinsically seeks to provoke instability and agonism in order to reveal the unequal power relations of space, firstly to the development practitioner as the assumed expert and outsider, but more provocatively to also reveal these relations to local inhabitants themselves (Hamdi, 2010a, pp. 71–72). His observations explicitly reference not only the need to facilitate and empower 'networks of solidarity, confidence and political entitlement', but crucially they must also enable 'the construction of alternative versions of their worlds'.

Contextualised against the complexity of informal settlements and communities of the Global South, Hamdi (like Turner before him) recognises that people themselves must assume control of their own futures and actively engage in producing their own spaces and relations (Burnell, 2012, pp. 135, 140–143). Such advocacy of intrinsic grass-roots control and freedom in development marks a direct connection to Turner's work over thirty years ago. As with Turner before him (1983, p. 7, 1976, p. 64), this is the central tenet of all Hamdi's writings: the observation that Westernised ideologies of what development should look and feel like are not compatible with what they actually can be.

Placed in the context of informal spaces and communities the possibilities of participation and disruption can thus offer vital political articulation of the realities, plausibility, and struggle for practical counter-hegemonic practices and catalysts for change (Mouffe, 2013, p. 139). Here disruption becomes not merely the opportunity to reveal hegemonic relations but also the potential to act as a social catalyst that demonstrates that development is not inevitable or hegemonic. It articulates the social opportunity and economic necessity to pursue development as change (Hamdi, 2010a, p. 166).

Thus, Massey's discourse provides a spatial contextualisation of the political discourse of Laclau and Mouffe in order to expose the political implications of a relational interpretation of space (2001, pp. 38–42). Yet her utilisation of Laclau and Mouffe's discourse remains decidedly retrospective and theoretical, and appears to not seek to offer examples of practical mechanisms for articulating and enacting alternative hegemonic space as acts of political agency. Yet Hamdi's advocacy of methodologies of small change provides such an intersection with Massey's advocacy of the positive disruption of hegemonic space and social relations. Such a comparison intentionally challenges assumptions of the informality of participatory development in the Global South as backwards and primitive, instead framing it as a methodological critique of the failures and limitations of Westernised space.

By drawing Hamdi's practices into comparison with what Laclau and Mouffe described as 'the moment of political articulation' (2001, p. 183), this chapter proposes that Hamdi uses disruption in order to reveal the hidden power geometries of spatial relations and, provocatively, suggest the potential to change them (2010b). This is where the notion of disruption as a catalyst becomes significant for both a practical and theoretical comparison of the potential not merely for development, but also for any potentially positive alternative and counter-hegemonic spatial relations in Westernised space.

Positive disruptions – a walking bus and a playground

At first glance Hamdi's development methodologies of agonistic disruptions appear relatively simple and unassuming. Yet they describe practices of creative exploration that look for interventions both to solve problems, but crucially also to generate social disruption and economic contestations. We have already explored both the theoretical trajectory and practical methodologies of participatory development that can and must be understood as disruption. Now, Hamdi's simple example of a project for a school *pied* (or walking) bus can be examined for its explicit agenda to reveal and contest local hegemonic relations.

As with so many examples explored in this research agenda, the apparent insignificance of a walking bus project belies the opportunity and implications of a far richer and more complex theoretical comparison. The project itself forms only one part of multi-stranded and explorative development agency (Hamdi, 2010a, pp. 119, 130). However, as an act of positive and agonistic disruption of cultural hegemony it is exemplary.

As part of a much broader discussion of community development practices that included engagements with agriculture, education, recycling, food, health, and political interventions, the walking bus was a singular response to a frequently recurring issue in informal and illegal settlements. This project worked with a community in a dense and informal settlement suffering from a variety of problems, differences, and disjunctions. However, Hamdi's skill was to observe and utilise their one collective common frustration – the area's lack of adequate roads and infrastructure meant it was difficult for children to get to school safely and there was no way to operate a traditional school bus system to the community.[3] This prompted Hamdi to advocate and quickly facilitate a relatively cheap and part-government-sponsored walking bus. Starting with a small number of such buses walking approximately forty minutes to and from the school, this project would serve both formal communal bus stops and direct pick-ups from isolated locations,

This in itself is a practical, efficient, and spatial articulation of innovative problem solving through small change. However, Hamdi is also explicit about the broader social disruption that this bus service would offer the community. In discussion with parents and the wider community, the bus would take different routes to school through neighbourhoods thought to be unfamiliar or risky by parents. Hamdi notes how 'It would be like a daily transect walk with children observing, recording, learning, informing. [...] It was a practical intervention with lots of potential for strategic planning' (Hamdi, 2010a, p. 113).

The walking bus would allow children to investigate different aspects of the neighbourhoods that traditional site analysis could not engage with, and specifically to engage with 'breaking down perceived borders between communities' (ibid.). These buses would cross borders of class, caste, and religion. They would confront the socio-spatial hegemony, implications, and expectations of the community by engaging with the universal desire for children to be given the best possible start in life. In the end these groups of school children would emerge as local area planning resources whose expertise could be applied to brainstorm ideas for improvements.

The spatial and cultural knowledge learned by the children would come to be used to inform future projects. The walking bus reflected the ability to build new spatial relations that tied the community together through the universal commonality of their commitment to educating their children. These practices became a mechanism to contest and disrupt both the local expectations of what development meant and the local authority's presumptions of the value of participation. The communities' original belief that the local authority had been letting down their children was disrupted. Instead they were empowered with renewed belief in their own political agency. Similarly, the local authority gained a renewed interest in alternative solutions for previously uncontested and unwanted problems.

Yet drawn from the wildly differing socio-economy and political context of development practice in the Global South, how can examples such as this can be compared to other spatial examples and practices from Westernised space?

In the UK, disruptive engagements with space can be observed in the work of Assemble Studio in London whose work includes the Cineroleum, Folly for a Flyover, Blackhorse Workshops, and the Yardhouse at the Olympic Park where their studio is now based. Yet perhaps their most socially provocative projects are engagements outside London and in contexts of relative economic absence: The Baltic Street Adventure Playground in Dalmarnock, East Glasgow, and the Cairn's Street Terraced Housing Renovations in Toxteth, Liverpool. In engaging in real critical social agency in the UK, Assemble Studio have explored methods that reflect the same social and political disruption of spatial relationships found in Hamdi's work: small changes that question big assumptions of architecture's role in the production of space.

These projects confront many of the same questions and issues of socio-spatial relations and the political agency of small change practices. Yet the wider potential for strategic change in such projects is rarely able to be followed through. They remain isolated by political and social entropy that seems to stifle architectural projects that attempt to exist outside conventional Westernised hegemonic relations.

The rarity of such examples of innovative social agency in UK architecture is telling, as is the 'London-centric' identity of the practices that engage in such agency. The accepted narrative of those practices that engage in innovative architectural agency is to evolve and grow from small scale and ephemeral projects with innovative social agency (which garner public attention), to larger (inherently more lucrative) projects and culturally accepted 'successes'. This model is now seemingly an un-written yet accepted path to follow for young and up-and-coming practices.

The question of the economic model by which practices like Assemble Studio or Studio Weave start up, exist, and grow remains unspoken in architectural discourse. The funding and economic context of London for such practices is yet to be balanced against their critical success. Similarly, their collective awareness of small change methodologies or the social sustainability of their practices needs further critical research.

Yet all of these questions coalesce against the daunting context of the meagre chances and opportunities that are available for socially and politically aware architectural students and practitioners to produce alternative spaces and architectures in the UK. How is it possible to support yourself economically whilst working in ways that are counter to the prevailing capitalist model of architecture and development in the UK?

Ultimately, the economic adversity facing anyone wishing to pursue such alternative architectural models must be questioned and confronted if alternative methodologies and critical social agency is to proliferate and become sustainable in Westernised space and the Global North.[4]

From space as stasis to space as multiplicity

In spite of the wider challenges of appropriating small change and disruption in Westernised space, Hamdi's practices in the Global South can be interpreted as explicitly seeking to question the hidden hegemonic social relations that form the complex spatial context and relations of development. Disruptive practices are thus also co-implicated with Massey's critiques of the hegemonic inevitability of development as an expression of structural interpretations of space (Hall et al., 2013; Massey, 2007, pp. 24, 211). Crucially, Massey specifically challenges Laclau and Mouffe's problematic insistence on 'the moment' of political articulation of counter-hegemony (or time) at the expense of an assumed passivity of space: 'For Laclau spatialisation is equivalent to hegemonisation: the production of an ideological closure, a picture of the essentially dislocated world as somehow coherent' (Massey, 2005, p. 25).

Here our analysis observes a recurring theme in the conjunction between space as representation, spatial relations, and hegemony as an ideological force. This analysis of space itself can be perceived as an extension of Lefebvre's spatial trialectic of the interdependence of lived, conceived, and perceived space. For Massey these links are assumed and inscribed within a way of perceiving and limiting our understanding of space which she describes as the 'the prison house synchrony of space and time' (2005, pp. 36–37). From this analytical departure Massey ventures further, citing Laclau's problematic reduction of space as merely the stasis representation of time:

> Any representation of a disclosure involves its spatialisation. The way to overcome the temporal, traumatic and unrepresentable nature of disloca-tion is to construct it as a moment in permanent structural relation with other moments, in which case the pure temporality of the 'event' is eliminated [...] in this spatial domesticisation of time.
>
> (Laclau, 1990, p. 72)

In response to this 'domesticisation of time' Massey advocates an understanding and utilisation of space in a profoundly different way. This alternative post-structural articulation explicitly identifies space and time as co-implicated partners in the constitution of the events and moments of political articulation (Massey, 2005, p. 158, 1996, pp. 116–117). This is a direct contestation of the historical, theoretical, and abstract equivalences of space as passive and static; a representation that Massey observes as having been constructed repeatedly by some of the greatest philosophers and theoreticians of the twentieth century: David Gross, Bruno Latour, Henri Bergson, Gilles Deleuze, Ernesto Laclau, and Michel de Certeau (2005, pp. 20–29). In contrast Massey advocates 'devel-oping a relational politics around this aspect of time-spaces' (2005, p. 180) that address their embedded and interdependent relations and geometries of power.

The central tenet of Massey's critique of space is the contestation of the restrictive and binary interpretations and articulations of space that defined structuralism, and that still survive within supposed post-structural thought. Her counter-proposition is for an alternative and political re-articulation of space as the sphere of something beyond mere representation of time as change (dell'Agnese, 2013, p. 116). This alternative conception of space as political, interdependent, and relational is crucial. It provokes the contestation of con-ventional conceptions of development and architecture which now cannot be articulated through formal planning and building alone. Instead, agents of spatial and architectural change must act to facilitate the contestation of social relations and practices, and here projects like Hamdi's walking school bus become exemplars.

Specifically building upon Laclau and Mouffe's proposition for 'radical democratic politics' and a pluralism of alternative (socialist) hegemonies (2001, p. xix), Massey's alternative understanding of change as implicated with space is explicitly a politicalisation of not merely space but the relations which produce it (2005, p. 65). Here our comparisons are reinforced by our reflections in previous chapters upon Lefebvre's dialectical interrogation of 'the reproduction of the relations of production' in *The Survival of Capitalism* (1976). Crucially we can again see Massey's positive articulation of space resonate with the wider theoretical context of this book. Her articulation of space as a relational product of agonism, difference, and change offers a critical lens to critique and contest the hegemonic constructions of space. This notion articulates a positive political potential of space that engages with the multiplicity, chance, and 'throwntogetherness' of social placemaking.

Thus, within all of the common misconceptions of space as representation, Massey is able to carefully and purposefully explicate the theoretical importance of the dynamics and relationality of space, by releasing it from mere stasis and representation (2005, p. 23). Her critique releases the synchrony of space and time as interlocked and purely representation and as closed systems of stasis. Massey notes that such stasis:

> robs 'the spatial' (when it is called such) of one of its potentially disruptive characteristics: precisely its juxtaposition, its happenstance arrangement-in-relation-to-each-other, of previously unconnected narratives/temporalities; its openness and its condition of always being made. It is this crucial characteristic of 'the spatial' which constitutes it as one of the vital moments in the production of those dislocations which are necessary to the existence of the political (and indeed the temporal).
>
> (Massey, 2005, p. 39)

This rejection of space as being inherently bound up in representation, ideological closure, and cohesion reveals the full limitation of Laclau and Mouffe's proposition for positive hegemony and agonism (Featherstone and Korf, 2012, pp. 664–665; Koczanowicz, 2011, p. 553). Their 'moment of political articulation' lacks the potential of space that Massey suggests comes from an appreciation of multiplicity. For Massey space is not stable, or coherent, or cohesive (2005, p. 116), but is instead the medium in which spatial relations, power, and change are socially produced. Space is inherently and necessarily chaotic and ridden with the consequences and implications of time and chance.

Massey's critique of both global and local space as bound by structuralism's closed systems and spatial relations can be critically linked to the implications of inevitability born from hegemony (2005, p. 59). By transcending the representation and inevitability of structuralism, this alternative and positive articulation of space is neither linear, nor fixed, but layered and overlapping, and because of these principles and characteristics, it is flooded with the possibilities found in multiplicity (Massey, 2005, p. 151, 1994, pp. 3–4).

Projected further and across assumed disciplinary boundaries this critically alternative proposition for dynamic, evolving, and necessarily incomplete relations of space can be compared to Hamdi's disruptions of the inevitability of development and to the alternative social relations, values, and spaces produced by communities of participatory practice and small changes. In keeping with our underlying pursuit of the positive alternative potential of space, Massey concurrently articulates the notion of interdependent space-time relations as framing the positive potential of space as a multiplicity:

> Space is as much a challenge as is time. Neither space nor place can provide a haven from the world. If time presents us with the opportunities of change and (as some would see it) the terror of death, then space presents us with the social and in the widest sense: the challenge of our constitutive interrelatedness; the radical contemporaneity of an ongoing multiplicity of others, human and non-human; and the ongoing and ever-specific project of the practices through which sociability is to be configured.
>
> (Massey, 2005, p. 195)

This interrelation of the instability of space, multiplicity, and development suggests further provocative implications at both local and global scales. Considered in the theoretical context of post-structuralism, Hamdi's methodologies and practices of socio-spatial disruption resonate with the specificity of space by focusing on local agendas and small changes. Such counter-hegemonic social practices engage in the specific social and material reality of informal contexts, learning from and articulating the positive potential that can be found in the minutiae of the everyday. For Massey it is this specificity and the 'throwntogetherness' of local place that can be used to articulate the political implications and potentials of humble changes (2005, p. 66). Massey's alternative advocacy of the interpretation of space as incomplete, relational, and specific reveals a framework and critical lens through which the potential of space might be revealed:

> The multiplicity and the chance of space here in the constitution of place provide (an element of) that inevitable contingency which underlies the necessity for the institution of the social and which, at a moment of antagonism, is revealed in particular fractures which pose the question of the political.
>
> (Massey, 2005, p. 151)

Space as such is understood as constituted of ever-shifting constellations. Multiplicity requires plural trajectories, connections, and relations which are intimately connected to the material and unfolding reality of local and global relations. Together, the positive multiplicity and chance of space must be recognised as inherently more challenging and complex in comparison to the abstract simplicity of structural thought and static definable space. Yet it is in the undecidability and chance of space that the positive political potential of small change and disruption can be found, explored, and articulated. Thus Laclau similarly notes the potential of space as found in contestation and subsequent relational openness: 'The moment of antagonism where the undecidable nature of the alternatives and their resolution through power relations becomes visible constitutes the field of the "political"' (Laclau, 1990, p. 35).

Specificity and undecidability

Having highlighted the intersection of the multiplicity and undecidability of space with the positive disruptions and small change methodologies of Hamdi's participatory development, this chapter can now project this comparison further into the theoretical discourse of the deconstruction of meaning and value. Here Massey is invaluable in providing a spatial turn with which to appropriate the premises of deconstruction and allow the positive tenets of Derrida's undecidability to be re-contexualised against the specificity of space and development practice.

This positive re-reading and analysis of Hamdi's disruptions of small change is further reinforced here by Massey's interrogation of the negative horizontality and dualism of deconstruction's structural articulation of space. Quoted here at length, Massey extensively critiques deconstruction as being appropriated and utilised in an inherently negative and dualistic formation:

> The focus is on rupture, dislocation, fragmentation and the co-constitution of identity/difference. Conceptualising things in this manner produces a relation to those who are other which is in fact endlessly the same. It is a relation of negativity, of distinguishing from. It conceives of heterogeneity in relation to internal disruption and incoherence rather than as a positive multiplicity. It is an imagination from the inside in. It reduces the potential for an appreciation of a positive multiplicity beyond the constant production of the binary Same/Other. [...] For, unavoidably, this imagination entails the postulation of a structure striving to be 'coherent' (in this very particular sense) but inevitably undermined by, or internally dependent upon, something defined as an 'Other'. This is the constitutive outside which is also the internal disruption. It is a way of thinking which posits identities (coherence) both in order to differentiate them counterpositionally one against the other (or, the Other) and in order subsequently to argue that they are, inevitably, internally disrupted anyway. What gets lost is coeval coexistence. [...] It is an imagination which, in spite of itself, starts from the 'One' and which constructs negatively both plurality and difference.
>
> (Massey, 2005, p. 51)

Within this critique Massey explicitly contests that deconstruction's negative utilisation of 'othering' sacrifices the social and political potential of plurality and multiplicity for an internal instability of post-structural identity, and thus fails to translate deconstruction beyond the linguistic and textual abstractions explored by Derrida (1987, p. 107). Yet by comparing these same ideas with her own advocacy of space as the sphere of multiplicity, the positive potentials of deconstruction in terms of values and spatial relations are materially re-contextualised to empower space as the sphere in which such political change must occur. Here the theoretical intersections drawn in chapter two between Lefebvre's and Massey's articulations of difference, appropriation, and multiplicity are renewed and projected further into positive utilisations of deconstruction as a pluralisation of social meaning, values, and purpose.

The terms and concepts introduced here are inherently complex, theoretical, and opaque, creating a seemingly vast gap to the reality of practical space outside.[5] Yet our overall research and critical comparisons reveal something new and different. These comparisons can positively frame Hamdi's catalysts and disruptions of small change as exploring real-world and practical articulations of the political implications of Massey's spatial contextualisation of deconstruction as 'an ever-moving generative spatio-temporal choreography' (2005, p. 54). The socio-spatial trajectories that Hamdi's practices create reveal alternative futures and potentialities of sustainable communities that are interpreted in our comparison as practical realisations of the need to 'shift in physical position, from an imagination of a textuality at which one looks, towards recognising one's place within continuous and multiple processes of emergence' (Massey, 2005, p. 54).

Such advocacy of 'undecidable' and 'textual' qualities of space reinforces Massey's recognition of space as the sphere that poses the question of the political. This question is realised in the everyday reality of the 'throwntogetherness' of living together (2005, pp. 140–141), and in the inherent chaos, risk, chance, disorder, and incoherence of space and multiplicity. Similar echoes are observable in Sennett's call to make positive use of disorder (1970), yet perhaps the clearest and most provocative theoretical description of the potential and necessity of instability are offered in Derrida's own articulation of deconstruction and the positive re-evaluation of chaos:

> This chaos and instability, which is fundamental, founding and irreducible, is at once naturally the worst against which we struggle with laws, rules, conventions, politics and provisional hegemony, but at the same time it is a chance, a chance to change, to destabilise. If there were continual stability there would be no need for politics, and it is to the extent that stability is not natural, essential or substantial, that politics exists and ethics is possible. Chaos is at once a risk and a chance.
>
> (Derrida, 1996, p. 84)

By articulating the positive implications of deconstruction outside the abstractions of its connections to language and meaning (Derrida, 1987, pp. 17–22)

Massey re-contextualises the concept within the political potential of space, and proposes something very different. In this analysis Massey crucially provides a spatial contextualisation of Derrida's advocacy of chance and chaos, arguing for the concurrent chaos and instability of space to be re-conceived as an inherently valuable facet of socially positive and politicised space. In contrast to structural space, these articulations of the necessary instability of space contest the uncomfortable connection between open and positive political space and how such space may be ordered, negotiated, and coded in specific places and social relations (Massey, 2005, p. 151).

This interpretation of space is not to suggest that formal space and strategic organisation become worthless or negative, but that the corollary informality, undecidability, and chance of rich space should be equally as valuable. However, for Massey what is missing is the articulation and embrace of the chaos of instability, and the valuing of disruption. Thus, spatial agents, development practitioners, and architects alike must learn to engage, articulate, and embrace their role in the process of socio-spatial disruption, and such open-ended practices must be re-valued as crucial to facilitating truly political space.

In the context of the deconstruction of space as offering positive undecidability, agonism, and political potential for change, Hamdi's practices once again can be re-read as providing invalidation, re-interpretation, and re-inscription of meaning and values in the most challenging of contexts. Just as was observed in Hamdi's walking bus example, positive space and multiplicity can contest the cultural hegemony and inevitability by deconstructing and confronting cultural assumptions of values and meaning. By understanding the social production of spatial relations as a means to contest and provoke change, the undecidability and chance of space engages with the creation of cultural meaning and values that define social space.

The implications of positive multiplicity and space to questions of difference, otherness, and values are explored extensively and speculatively in chapter five. Yet it has been necessary to contest the foundations of deconstruction here in order to reveal the complex implications of releasing space from hegemony and static representation. If, as observed in Hamdi's practices, space is to be open to positive change and agonism then in Massey's critical discourse this chapter observes a framework that is foundational to any conception of development as open and free, and to any attempts to re-imagine architecture and space as relationally constructed positive social processes.

These notions connect with Hamdi's open methodologies of practice as acts of exchange and learning between partners. Massey's utilisation of both Laclau and Mouffe's advocacy of the political necessity of positive agonistic political theory and Derrida's notion of deconstruction and the positive undecidability of meaning and values can be critically re-read and compared as part of an advocacy of the positivity of space through disruption:

> From deconstruction, the notion of undecidability has been crucial. If, as shown in the work of Derrida, undecidables permeate the field which had

previously been seen as governed by structural determination, one can see hegemony as a theory of the decision taken in an undecidable terrain.

(Laclau and Mouffe, 2001, p. xi)

Thus Massey's re-articulation of the politics of space as innately part of the sphere of multiplicity and political negotiation of social relations draws equally upon Laclau and Mouffe's reading of hegemony, and Derrida's deconstruction and undecidability. In the process Massey's discourse has provided various opportunities for comparison with Hamdi's open-ended participatory practices of disruption. Hamdi's methodologies begin to represent more than mere practical expressions of inevitable development, but instead show an engagement with the 'undecidable terrain' of development. This connection and comparison provides a crucial spatial link between the open positivity of multiplicity as a projection of deconstruction, and the practical reality and methodologies offered by Hamdi's participatory practices.

Re-imagining space and architecture via pickle jars

Massey's positive articulation of undecidability as concurrent with the multiplicity of space becomes important as a means to connect the unstable and undecidable nature of place and context, and the spatial specificity that informs local negotiations of politics (2004, p. 11). If the chance and chaos of deconstruction are bound to instability and incoherence, then changing political space must always remain an incomplete and ongoing practice, or in more conventional terms, a dialectic. This implies a necessity for political space to be practised in order that it retains dialectic instability. Yet the positive openness and unknowing of space remain seemingly intractable and implausible when set against the hegemonic passivity of the Global North (Harvey, 2012, pp. 24–25). Thus, when compared and contextualised against Hamdi's participatory development practice, his methodologies of local disruption and advocacy of social change can be observed as producing alternative spatial relations that are negotiated and developed from within a truly undecidable terrain of participatory practice (Hamdi, 2004, p. 127).

The fact that such positive political space must always remain unstable and undecidable can now be perceived as the implication of truly democratic and political space. In this context, our critical premise remains that Massey and Hamdi offer intersecting practical and theoretical confrontations with capitalist spaces of inequality and hegemony. In their own ways they each advocate the agonistic negotiation of relations of local and global specificity within space. If change is generated by instability, then the shared social act of negotiating space represents the possibility of a truly open and free politics of space. The challenge for all spatial advocates, agents, and participants is having the confidence to treat space in this way in the face of the implications such a contestation supposes: 'Instead of trying to erase the traces of power and exclusion, democratic politics requires that they be brought to the fore, making them visible so that they can enter the terrain of contestation' (Mouffe, 1993, p. 149).

In this context, Hamdi's disruptions and catalysts can be suggested as revealing such 'traces of power and exclusion' and reconstituting space with the potential of politics and alternative possibilities. Where Massey advocates a theoretical resilience found in the notion of space as the sphere of multiplicity and chance, Hamdi's development practices can be seen as actual realisations of counter-hegemonic political space. Here the resonance between Massey's theoretical advocacies of positive counter-hegemonic practices and alternative social relations, and Hamdi's project of facilitating a local community vegetable pickling industry is exemplary.

The home pickling project is one of several wonderfully simple and imaginative examples of participatory development practice that Hamdi uses to demonstrate the implications of his methodologies for engaging informal settlements. The methodologies and practices that define this example are relevant to this discussion as they represent the key characteristics of Hamdi's articulation of development practice as a positive and catalytic disruption. The simplicity and humility of these practices are too easily overlooked or undervalued, yet are highlighted and celebrated here:

> we encountered one enterprise, easy to miss, the smallest I have seen, along one of the many hidden pathways leading to the Centre. Two glass jars. [...] The jars contained five pickled cucumbers each, which were for sale to passers-by.
>
> (Hamdi, 2004, p. 85)

This simple social relation of pickling vegetables would be remarkably easy to overlook without a pronounced determination to contest expectations and hegemony. Yet Hamdi's informal encounter and creative humility regarding this discussion of pickles reflects a contestation of truly open and disruptive practice, and a subversion of both local and global expectations of development ideology. Emblematic of small change, it starts with the very small idea of facilitating already existing social relations by supporting and expanding the local enterprise in pickle jars. In this act Hamdi instigates a humble and almost unbelievably simple disruption. But it carries with it far greater implications than at first appear. And crucially it offered a concrete realisation of real and meaningful change made possible.

Local people had already tried to make this idea of pickling and selling vegetables into a more substantial enterprise and yet they had failed to make it into a viable business (Hamdi, 2004, p. 86). The assumed explanation for this was merely that this simple cottage industry was nothing more than a couple of women making pickles from subsistence gardening. The reality of their economic context based upon a capitalist hegemony of inequality and imbalanced power geometries of formal economies seemingly invalidated their enterprise and hard work as a source of inspiration for development.

Hamdi's recognition of the greater opportunity of this enterprise runs counter to the spatial relations of capitalist hegemony that had isolated this enterprise. The previous project had failed to succeed not because it wasn't socially viable and potentially economically beneficial, but because it didn't conform to accepted formal models of development such as economies of scale, profit margins, and so on.

Whilst this would suggest that the project had apparently already been shown to be non-viable, this is exactly where Hamdi decided to enact a form of participatory practice as disruption. This alternative interpretation of the pickle jar project in comparison with Massey's space of multiplicity and undecidability suggests that if this enterprise did not have to be constrained by the spatial relationships of capitalist hegemony, it could potentially contest and transcend the inevitability of hegemonic development. In attempting to facilitate and connect this pickling enterprise, Hamdi enacts a disruption to the local economic hegemony and provokes a small, targeted, and agonistic act of counter-hegemony within the vast problems of this informal community.

Instead of being constrained by hegemonic assumptions of development Hamdi acts to advocate and facilitate an emerging network of alternative social relations that connect pickles to complex and specific models of socially sustainable enterprise. This allows him to connect the simple act of pickling to a wider context of social relations that included school reform, ecology, education, food, and helping with malnutrition (Hamdi, 2004, p. 87). Considered in relation to our wider critical theoretical framework, this can be considered as an alternative social hegemony where spatial relations are working towards something other than capitalist relations. Whilst this project created employment and enterprise, it did not do so merely for the profits of a few people, but looked to articulate new alternative and positive social relations that reflected realisable, scaleable, and distributed economies of sustainable enterprise.

Here, development practices of small change and a cottage industry of pickling vegetables describe a political articulation in a space and moment of change. This is change built upon necessity, but then this necessity is not translated into Western capitalism, but instead produces new and alternative socially sustainable relationships. As a product of economic and social necessity this is recognition that Western standards, models, and forms of development cannot and must not be adhered to as expectations for the world.

> It served as a catalyst: a community of interest energising around a common need. And later [...] they would be welcomed across entrenched social boundaries. Where once there were barriers – a place to hide, in the face of the threat from others, from evictions, the low self-esteem imposed by poverty or the real threat of class conflict – these new boundaries offered a sense of belonging and connectedness. They offered a common context of meaning where individuals acquire identities as members of a larger social network [where] the network generates its own boundaries. [...] This, then, is the 'soft city' of dreams, expectations, interests held in

common and webs of relationships, not easy to explain or model because
its structure is largely invisible and, in any case, always changing.

(Hamdi, 2004, p. 88)

Whereas Massey provides a theoretical explication of positive space,
Hamdi achieves the same thing practically by simply supporting the idea,
endeavour, and social sustainability of growing and selling pickles (Hamdi,
2004, p. 87). Considered as a social catalyst, Hamdi is clear that the act of
facilitating a catalyst itself was more important than whatever outcome might
have come from pickles. Such acts suggest to local people that in even the
smallest activity there are alternatives and possibilities that exist outside cultural hegemony and its cohesion and inevitabilities. Such a small endeavour
might not seem to suggest a great achievement in the development of impoverished informal settlements, yet with pickles and imagination Hamdi is
advocating an understanding of development values that aren't constrained
by inevitabilities of Western cultural hegemony.

This practice describes an agonistic social agency within space that seeks to
articulate new relations between social and political institutions and to see the
world in a different way. Providing people with alternative aspirations and
expectations can begin with pickles and go on to mean something far more
important. Thus, because this project was shown to be possible, it suggested
that the local economic and material reality wasn't fixed and inevitable. The
community was free to become something more in time (and space), and was
shown that it was possible to strive towards a better future:

It was all at once ambitious and imaginable. What if, we asked, these
same organisations became partners in education for sustainable development, an alliance of local, national and global institutions in the governance
of education? Who would win and who might lose out? What might
happen to Tandia and her colleagues if big money and big organisations
got involved? We decided to get it all going first in a small way, without
outside help and later, maybe, involve others when we were ready to scale up.

(Hamdi, 2004, p. 88)

The pickle project articulated a positive political moment through a spatial
practice of specificity and political agonism that contested a challenge to
inevitability. The small disruption of a community enterprise of pickle growing
might change an entire city, in time. Or it might not. What was important
was simply the potential that it could and might generate change and that
local people (ambitious residents, local politicians, and agencies) would see
that alternatives were possible. They would talk about pickles, and about their
lives and aspirations and would begin to realise that capitalist spatial relations
are not inevitable and not the only solution.

Here the intersection with Massey's theories of the positive chance and
undecidability of space are subtle yet profound. Crucially this desire to see

things differently comes not at the point of an ideological imperative but upon the necessities and specificity of a place that exists at the peripheral edges of Western economic hegemony. In this context Hamdi recognises the necessity of alternative ways of doing things: of seeking to produce alternative social and economic relationships, and alternative values. Whilst the connection of this example to our wider implicit commentary on Westernised space and architecture is on the surface difficult to see, if it can be allowed to stand as an exemplar of counter-hegemony then its implications might begin to appear quite profound.

If Westernised architecture and space were considered with both the humility of aspirations and respect for local social and material contexts as Hamdi advocates, then its potential to engage with people and space would be greatly increased. An architecture that sought to contest hegemony would not have to do so through structure and form, but could seek to do so through socio-spatial practices of empowerment and facilitation. This would not be an architecture based upon form, style, and taste, but an engagement with the space and values of undecidable terrain and the articulation of positive alternative social relations and space through disruption and catalysts of change.

Catalysts and scaleability

The comparisons between theoretical and practical articulations of the positive multiplicity of space that have been highlighted in this chapter are vital. Where these discourses intersect they reveal the potential for a disruption and catalytic emergence of alternative social relations as catalysts from within the specificity of place.

Yet crucially these spatial practices do not reflect attempts to provide a fixed answer and solution to space. Hamdi's practices are explicitly intended to act as catalysts that test the water of complicated and apparently cohesive spaces. Furthermore, Hamdi is explicit in this regard, noting that the disruption and social catalytic effects of small change are intended to 'enable outsiders to focus their efforts where need is greatest and together to search for triggers for change' (2004, p. 96).

Hamdi's spatial agency of disruption generates this potential by acting as a catalyst that holds within it the potential for scaleable social change, be that in the form of pickle jars, bus stops, water taps, or waste recycling programmes. These catalysts are not fixed futures. Each of them is only ever an aspiration, a challenge to the inevitable and a possibility of alternatives and change. Catalysts such as the pickle project offer a point of departure in participatory development practice that allows Hamdi to articulate the role of the practitioner as a facilitator of alternatives and possibilities: 'It offers a different process and, at the same time, consolidates the role of the outsider as a catalyst, mediator, facilitator or enabler' (Hamdi, 2004, p. 105).

Catalysts generate discussion, argument, disagreement, and the potential to produce agonistic social space and the balance of truly political space. Hamdi advocates these catalysts of small change to create forward momentum and binds them into notions of dreams, relations, networks, boundaries, belonging, and connectedness. The subsequent status of development agency as a facilitator and enabler of open change contests both the inevitability of development and also the identity of 'outsiders' and 'recipients' of such projects (Hamdi, 2004, p. 128).

Such a model of development does not prescribe that there must indeed be change, or the form that change may take, it merely creates the space needed for the potential of change and alternative stories to be a possibility (Hamdi, 2004, p. xvi). Space is 'always changing' and is a part of the 'common contexts of meaning' (Massey, 2009, pp. 25–26); here we can explicitly link Hamdi's practices to Massey's articulation and interpretation of space as the sphere of possibilities, and as existing as a medium of positive multiplicity, relationality, and specificity. Thus, Hamdi's agonistic, disruptive, and catalytic practices are always:

> searching for ways to join people and organisations together, build ties in some circumstances and loosen ties in others, expand the scale of small initiatives, open doors to ideas, to other people, to organisations who can help find money and enterprise, reframe questions, legitimise and give stays. And also to be rigorous, flexible and principled, working sometimes with individuals for the collective good and not always with communities. And, importantly, as one goes about one's work, learning that sometimes it may be best just to leave things alone.
>
> (Hamdi, 2004, p. 85)

This comparison highlights a further significance of Hamdi's methodologies to the importance of actions, disruptions, and catalysts that are not predicated upon knowing what their outcome will be or whether they will succeed. They are not implicated or interdependent with a hierarchical political ideology, instead empowering networked governance and grass-roots radical democracy (Mouffe, 1999, pp. 753–755).

Catalysts are the starting point. They are the moment of intersection and transgression where new social relations and practices are created and contested. They are never intended as a resolution in themselves, merely facilitating moments and spaces of political articulation through participatory practices of negotiation. This insight into Hamdi's methodologies of practice reflects the explicit importance of practising without knowing or prescribing an answer or even necessarily a specific problem:

> We worked somewhere between knowing and not knowing what might happen. We provided ample opportunity for the results of our first decision – routing the bus line, positioning the bus stop – to tell us something about subsequent actions that may induce a change of mind, a change in direction or even change of objective. We avoided pre-emptive answers, in

this case to community, and instead facilitated its emergence. [...] We see in this way of working, a kind of practice that does not rely for its effectiveness on certainties or complete information. [...] Improvisation then becomes a means of devising solutions to solve problems which cannot be predicted, a process full of inventive surprises that characterise the informal way in which many poor people gain employment, make money and build houses.

(Hamdi, 2004, p. 98)

In many respects this suggests that the ultimate outcome is not important, or at least that the outcome should be unknown, or more profoundly that there can be no fixed outcome if such spaces are to be truly political. This is once again the disruption of the inevitability of hegemony – whether it be capitalist, Westernised, or merely unequal space, the implications remain the same (Lummis, 1997, pp. 64–65). Thus, it is the practical imperative of not knowing the 'end result' or 'resolution' of development practices that becomes of the utmost importance. Here once again we can draw a comparison with Massey's spatial articulation of Derrida's values of otherness and the undecidable nature of politically positive space.

Hamdi explicitly notes that practices based upon not knowing or assuming answers are profoundly uncomfortable for old-paradigm thinking and the traditional assumptions and concepts of development (Hamdi, 2004, p. 13; Long, 1992, p. 270). Conventional development practice is constructed around clear demarcations of quantifiable results and policy-based planning whose goal must be to find the answer to the problem of space by the quickest and most structurally efficient means. As such, practising without knowing this answer or even without the ambition of achieving an answer is challenging. It is also profoundly provocative, and is an aspect of Hamdi's practice that is much overlooked and underestimated. In light of this, this chapter suggests that further insight is offered in Hamdi's recollection that: 'It is about getting it right for now and at the same time being tactical and strategic about later' (Hamdi, 2004, p. xix).

This articulation of practising without knowing what the end result might be suggests a provocative contestation of architecture and design that might easily be interpreted again as just a simplistic approach for intervening in contexts on the economic and social edge of necessity. Yet by again drawing these methodologies into comparison with the far broader theoretical context of post-structural space and deconstruction we can see in Hamdi a practical realisation of Massey's positive and spatial contextualisation of an 'ever-moving generative spatio-temporal choreography' (2005, p. 54).

Once again the wider implications of this comparison resonate with the wider premise of our study: to re-imagine space and architecture in the context of difference, multiplicity, and otherness. If Hamdi's open-ended practices can now be read as a contestation of positive counter-hegemonic social relations and politics, then whilst the examples of pickle jars and walking buses remain perhaps only emblematic, the underlying principles of openness and critical engagement with space as a multiplicity of social relations begin to offer valuable opportunities for critical reflection.

In these practices Hamdi explicitly values the ambiguity, and the shifting and open nature of participatory catalyst projects, advocating socio-spatial practices that provoke critical and agonistic dialogues. Such open-ended practices are intentionally started as small, realistic, and graspable actions that involve and engage people in the negotiation of space. Crucially, the implicit undecidability of these practices is explicitly not the replacement of one hegemonic imbalance with another. These practices merely facilitate and release other people to imagine and try out other alternatives (Hamdi, 2010a, p. 93). Thus, the challenge of such openness to truly post-structural and plural space is perhaps reflective of why Westernised space remains an expression of structuralist hegemony.

Interdependent with the commitment to localised and efficient small-scale changes, Hamdi's alternative imaginations of space and development are implemented with the specific intention of 'going to scale' (2010a, pp. 67–68). This commitment to strategic change is integral to the social sustainability that Hamdi's practices embody, whilst also suggesting the far greater political potential such scaleable social change might contain within it. The scaleability offered by such alternative forms of grass-roots social practice are in critical contrast to the traditional hierarchical models of political authority. Participation and social enterprise act as catalysts for social relations founded upon material and political reality, with the implication that such forms of network governance and grass-roots politics offer potentially dramatic socio-cultural change. They contest and confront the question of who governs cities, and the purpose and values that such governance produces (Yates, 1980). Thus as Shann Turnbull observes: 'Currently, we seem to face a choice between state-run enterprise or state regulation, or privatised and public interest companies. Stakeholder governance provides an alternative' (2002, p. 32).

This alternative notion of 'stakeholder governance' is inherently generated by the ability of catalytic development to facilitate a grass-roots network of resilience and enterprise, as observed by Hamdi in the Global South. It provides a resonance with Turner's earlier work on progressive housing and autogestion to facilitate the alternative and positive leveraging of social capital. In contrast to the Western urban context of neoliberal capitalist inevitability, Hamdi's catalysts explicitly act to recognise, value, and facilitate the scaleability of hidden and suppressed subaltern livelihoods. The agency of grass-roots networks and stakeholder governance articulates alternative trajectories for development that exist within and emerge from the cracks of hegemonic space. They only require the positive spatial agency of facilitation and development to be set free to contest contemporary capitalist space:

> Practice sparks the process by which small organisations, events and activities can be scaled up. This can happen in various ways: quantitatively,

where programs get bigger in size and money; functionally, through integration with other programs and other organisations both formal and informal; politically, where programs and communities can wield power and can become part of the governance of cities; and organisationally, where the capacity to be active increases and becomes sophisticated and influential – at which point it becomes a higher order of organisation. Emergents and going to scale are, therefore, complementary processes: practice is a catalyst to both.

(Hamdi, 2004, p. xix)

Hamdi's philosophy of catalytic practice and acting in order to induce others to act is a cultivation of the necessary environment for change from within.[6] It starts on the ground from small beginnings that have emergent and scaleable potential to induce social enterprise and change. This connection of catalysts to strategic scale is not a rejection of municipal authority and governance, but a cultivation of an alternative grass-roots collectivism of self-organising power. It is not a rejection of social hegemony and infrastructure, but a proposition for a social and material alternative of plural and networked multiplicity. Thus in spite of the inherent challenge that such provocative counter-hegemony implies, Hamdi remains committed to the positive implications of this small change:

Not all small beginnings achieve strategic value. Indeed most times, strategic change is hard to come by – the filter upwards of ideas and learning clogs with those who will resist change and those with old-style laws and regulations left over from days of old-paradigm thinking. The connectedness it all demands between events and organisations doesn't happen because people are still dependent, or because they have only recently won their independence and are not yet ready to move to inter-dependence. But none of this diminishes the importance of the effort and the gains on the ground.

(Hamdi, 2004, p. 90)

The fact that such alternative practices are yet to become connected to global relations and socio-cultural expectations of space only reflects the ideological impenetrability with which neoliberal capitalism continues to subsume social and political frameworks of space. Hamdi's and Turner's challenges to these relations in the Global South merely highlight the global inequality and implausibility of neoliberal economics, and the frustrating gap between necessity and want, value and excess. Hamdi's methodological success taking disruption and catalysts from small change through to strategic big change only serves to highlight the inability of alternative architecture in Westernised contexts to 'go to scale'. Re-imagining space and architecture via small-scale disruptions and catalysts requires a confrontation with and inversion of conventional short-term capitalist economics and neoliberal policies.

Global and local – relational interdependence

The opportunities to disrupt and change space with catalysts is palpable in both Hamdi's and Massey's work. The local and political specificity of these changes may superficially appear to suggest an agenda of resistance and reaction against outside forces. Yet in contrast to any reductive identification of localism and 'geographies of resistance' (Massey, 2004, pp. 10–12) implied by the scale, specificity, and undecidability of space, for Massey local space is always, and has always been, inherently implicated in the production of the global. Any such traditional calls of localism as resistance to global relations and power sacrifices the political potential of truly open and relational space, losing the potential points of purchase it offers (Massey, 2004, pp. 14–15). Thus, Massey's critical contextualisation of global and local in terms of multiplicity and relationality necessitates a final thread of comparison with the inherent aspirations of scaleable change observed in Hamdi's positive participatory projects.

For Massey, local space exists interdependently with practices and processes in relational space-time (1999, p. 33). Contrary to calls to nationalism and cultural specificity, local space is constructed out of a multiplicity of trajectories of space and is inherently reliant on an openness to chance and change (Massey, 2005, pp. 64–65). Each local space and specific context is continually being produced by its local and global connections, and as such is shifting and contracting in response to its economic, social, and political relations. Within the complexity of such global and local space-time the true political implications of such openness is found in the terms in which the power-geometries of relations are constructed (Massey, 1994, p. 5).

As such, attempts to develop a practical relational politics of such time-spaces forces a confrontation with the specific, interlocking, and embedded geometries of power. Based upon this critical analysis, this chapter further contends that Hamdi's practices of revelation, disruption, and explicitly scaleable catalysts can be compared as practical contestations of such time-spaces. This comparison is born out in Massey's description of a relational politics of place as involving:

> both the inevitable negotiations presented by throwntogetherness and a politics of the terms of openness and closure. But a global sense of place evokes another geography of politics too: that which looks outwards to address the wider spatialities of the relations of their construction. It raises the question of a politics of connectivity.
>
> (Massey, 2005, p. 181)

Such a contestation of the open and connective relationality of the local and global is a direct theoretical reflection of Hamdi's advocacy of both locally targeted and yet specifically 'scaleable catalysts' for sustainable social enterprise and development (2010a, p. 139, 2004, pp. 110–115). It rejects the simple binary surface of local global relations in favour of the political potential of relational agonism. In accepting globalisation as being an intrinsic condition of space-time the questions for Massey become: what kinds of alternative interrelation are allowed to underpin development, and what is the nature of such a political project?

The inherent critique of political utilisation of hegemony and counter-hegemonic identities in local and global agendas intersects with broader critical discourse of universal political struggle and cultural uniqueness. The most recent and controversial indictment of such discourse is Vivek Chibber's *Postcolonial Theory and the Specter of Capital*, which has provocatively contested the intellectual and subsequent political implications of discourses of cultural subalterneity as subverting the underlying imperatives of universal class struggle (2013, pp. 217–218). Whilst rebuttals to this discourse are only beginning to surface (Robbins, 2013), this contestation of uniqueness and specificity might already be partially articulated in Massey's global–local relationality and specificity.

The implications of local and global trajectories of multiplicity in relational space suggest very different geometries of power. Within such a context each contestation and struggle for change and difference is an extension and meeting along lines of constructed equivalence and relational equality. The global and local are interdependent and interconnected – all of our actions and agencies are interconnected. The practice and process of negotiating and engaging in the contestation of relational topographies of power offer an imagination in which local struggles are relationally independent with global common struggles against hegemonic cohesion (Massey, 2005, p. 182).

This understanding and contestation of the relationality of global and local space provokes an immensely complicated articulation of space, but one that still remains grounded in grass-roots social relations and practices. These are exactly the types of relation that this chapter observes Hamdi contesting and engaging with, and thus this comparison is reinforced by Massey's observations of the practical challenge of such a confrontation:

> One effect is to demand far more of the agents of local struggle in the construction of both identity and politics than there is room for in the topography where identity seemingly emerges from the soil. Theorists of radical democracy, on the other hand, have rarely engaged with the complexity and real difficulty of this construction of equivalences.
>
> (Massey, 2005, p. 182)

When re-contextualised in this way disruption, catalysts, and ambition of scale in Hamdi's development practice methodologies are critically comparable to Massey's advocacy of the multiplicity of space. This chapter has shown that they provide a similar fundamental political agenda to space as was previously discussed in Lefebvre's narrower theoretical articulation of 'the right to difference', and suggest a form of counter-hegemonic practice akin to Mouffe and Laclau's advocacy of positive social agonism in political space (Mouffe, 2013, pp. 17–19).

Yet unlike chapter two's theoretical discussion, the specific comparison of practical methodologies in this chapter contests space as the sphere and field of multiplicity in practical small change of political engagement. It is this contextual comparison that makes key examples like Hamdi's articulations of alternative development so spatially provocative, and valuable in our attempts to pursue a re-imagination of Westernised architecture and space.

Small change and positive multiplicity

Through our critical comparisons and close re-readings of key theoretical concepts Hamdi's methodologies of participation and disruption have been re-appropriated and re-framed, offering far more positive spatial implications than they might suggest on first inspection. The critical reflection and comparison of the observations and practices of participation undertaken by Hamdi critically articulates the political and social implications and specificity of practices of small change in a post-colonial and globalised context of multiplicity.

The social and political disruptions generated by Hamdi's practices in contexts at the periphery of economic instability are thus recognised as realisations of dialectical social change through the interrogation, disruption, and production of alternative sustainable social relationships. In this comparison, Hamdi's disruptions of small change can be valued as unique practical articulations of the potential of space to produce positive counter-hegemony and spatial relations of multiplicity. When drawn into this comparison these disarmingly simple practices have been shown to reflect common spatial aspirations with the pioneering spatial and political theory of Massey et al.:

> Conflict and division, in our view, are neither disturbances that unfortunately cannot be eliminated nor empirical impediments that render impossible the full realisation of a harmony that we cannot attain because we will never be able to leave our particularities completely aside in order to act in accordance with our rational self – a harmony which should nonetheless constitute the ideal towards which we strive. Indeed, we maintain that without conflict and division, a pluralist democratic politics would be impossible. To believe that a final resolution of conflicts is possible – even if it is seen as an asymptotic approach to the regulative idea of a rational consensus – far from providing the necessary horizon for the democratic project, is to put it at risk.
>
> (Laclau and Mouffe, 2001, p. xviii)

As purely theoretical propositions these abstract articulations remain frustratingly unrecognisable, particularly when faced with the extreme cohesion and oppression of Western hegemonic space. Yet in critical comparison to Hamdi's practices the theoretical discourse of Massey's relational space is made tangible in simple practices of small change that revel in the political potential for change and the chaotic chance of space. Thus the comparison to Hamdi is illuminating, practically, economically, and geographically as it places the proposition for change at the informal and alternative periphery of global society. At this periphery we find the instability and undecidability that is needed to invoke positive change. Here we find the conflict and otherness that is missing in Western contexts:

> Change requires interaction. Interaction, including of internal multiplicities, is essential to the generation of temporality. Indeed, were we to assume the unfolding of an essentialist identity the terms of change would be already given in the initial conditions. The future would not be open in that sense. And for there to be interaction there must be discrete multiplicity; and for there to be (such a form of) multiplicity there must be space. [...] We cannot 'become', in other words, without others. And it is space that provides the necessary condition for that possibility.
>
> (Massey, 2005, p. 55)

This final moment of comparison in this chapter is found in this notion of alternative spatial practice as engaging with openness, unknowing, and otherness. Such practices of social agency are found in the theoretical provocations of instability, incoherence, and agonism, all of which are violently uncomfortable relative to Western existence. As evidenced by Hamdi, disruption is necessary even in informal settlements, where capitalism has no viable claim to logical coherence. This suggests that the potential for political spaces of agonism in the Global North will either not come at all, or perhaps only ever in peripheral contexts of economic necessity.

Or perhaps not. Perhaps there is still something more to be found in the disruption of small changes. Perhaps within the idea of small changes by communities of practice there is retained enough scope for alternative ways of thinking and acting. The relations between things can be changed. Even within the hegemonic and politically passive space of neoliberal capitalism, perhaps in small things like pickle jars we might still find positive spaces for positive actions. Perhaps, even within the hegemonic realms of the Global North, there remain possibilities for porosity, for power's leakage or a scaling up of spatial and social disruptions to a point of catalysing major changes.

Thus, our re-contextualisation of critical themes within development discourse and practice creates precisely the critical opening implied by Massey's desire for multiplicity, relational global development. It is this opportunity to further contextualise and re-examine Massey's foundation of positive spaces of multiplicity and difference that will be pursued in the remaining chapters of this book.

Similarly, Hamdi's discourses and practices must also be reconsidered in light of their intersection with such broad strands of critical spatial theory. Once again, it is this opportunity to further contextualise and contest the global, intellectual, and theoretical value of Hamdi's practices that will be pursued in the next chapters.

Notes

1 Once again, here we will explicitly reiterate that the agenda of our research is not to glorify or voyeuristically fetishise the informal spaces of the Global South as objects, but to celebrate the positive potential they represent that people and space have for change and innovation (Dovey and King, 2013; Hatherley, 2014).

2 Coincidentally, whilst writing this chapter it has been confirmed that the whole Docklands development site has been sold to the Qatari national investment fund. The global implications of such spatial relations are clear, and reflect distinct power geometries with respect to local, national, and global relations. Not long afterwards HSBC have also begun to question the continued location of their headquarters in London, citing the political changes and legal restrictions placed on London after the financial crisis. The truly global relational construction of such space is writ large in such moments.

3 Intriguingly Hamdi's practice in this instance borrows from a successful walking bus system in Lecco, Italy, in which the local authorities hire bus drivers who walk the children to school rather than drive them.

4 This is research that I ardently seek to pursue in future work, but must fall outside of the remit of this book which must exist as a foundation and framework for such future research. It is a far-reaching set of questions with the potential to challenge the very basis of the architectural profession, its authority, education, and role in society.

5 The challenge of critically appropriating deconstruction into architectural theory and practice is highlighted by the various ludicrous attempts of contemporary architects who have managed to translate complex social and political theory into buildings as mere expressions of self-referential form and aesthetic.

6 Once again reflecting a link to Marx's, Lefebvre's and Harvey's advocacies of the dialectical opportunities for space to change us, but reciprocally, for us to change space. See chapter one.

References

Baan, I., Brillembourg, A., Eidgenössische Technische Hochschule, Department Architektur, Institut für Geschichte und Theorie der Architektur, 2012. *Torre David: Informal Vertical Communities*. Müller, Zürich.

Barnett, C., 2004. Deconstructing Radical Democracy: Articulation, Representation, and Being-With-Others. *Political Geography* 23, 503–528.

Bauman, Z., 2000. *Globalization: The Human Consequences*. Columbia University Press, New York.

Boano, C., 2013. Architecture must be Defended: Informality and the Agency of Space. *Open Democracy*, 24 April. Available at https://www.opendemocracy.net/op ensecurity/camillo-boano/architecture-must-be-defended-informality-and-agency-of-space (accessed 9 March 2016).

Bower, R., 2016. Doreen Massey's Spaces of Multiplicity, Relationality, and Specificity, in: *Architecture of Alterity*, Symposium, held at University of Edinburgh, 25–27 May 2015. Cambridge Scholars, Cambridge.

Burgess, R., Carmona, M., Kolstee, T., 1997. Contemporary Spatial Strategies and Urban Policies in Developing Countries: A Critical Review, in: Burgess, R., Carmona, M., Kolstee, T. (Eds), *The Challenge of Sustainable Cities*. Zed Books, London.

Burnell, J., 2012. Small Change: Understanding Cultural Action as a Resource for Unlocking Assets and Building Resilience in Communities. *Community Development Journal* 48, 134–150.

Capra, F., 2002. *The Hidden Connections. A Science for Sustainable Living.* Harper Collins, London.

Chibber, V., 2013. *Postcolonial Theory and the Specter of Capital.* Verso, London.

Clarke, J., 1991. *New Times and Old Enemies: Essays of Cultural Studies and America.* Harper Collins, London.

Cornwall, A., 2004. Spaces for Transformation? Reflections on Issues of Power and Difference in Participation in Development, in: Hickey, S., Mohan, G. (Eds), *Participation: From Tyranny to Transformation.* Zed Books, London.

de Graaf, R., 2015. Architecture is Now a Tool of Capital, Complicit in a Purpose Antithetical to its Social Mission. *Architectural Review,* 24 April.

dell'Agnese, E., 2013. The Political Challenge of Relational Territory, in: Featherstone, D., Painter, J. (Eds), *Spatial Politics: Essays for Doreen Massey.* Wiley-Blackwell, Chichester.

Derrida, J., 1987. *Positions,* 2nd Edition. Athlone Press, London.

Derrida, J., 1996. Remarks on Deconstruction and Pragmatism, in: Mouffe, C. (Ed.), *Deconstruction and Pragmatism.* Routledge, London.

Dovey, K., King, R., 2013. Interstitial Metamorphoses: Informal Urbanism and the Tourist Gaze. *Environment and Planning D: Society and Space* 31, 1022–1040.

Esteva, G., Prakash, M.S., 1998. *Grassroots Post-Modernism: Remaking the Soil of Cultures.* Zed Books, London.

Featherstone, D., Korf, B., 2012. Introduction: Space, Contestation and the Political. *Geoforum* 43, 663–668.

Fraser, M., 2012. The Future is Unwritten: Global Culture, Identity and Economy. *Architectural Design* 82, 60–65.

Gramsci, A., 1971. *Selections from the Prison Notebooks.* Lawrence and Wishart, London.

Hall, S., Massey, D., Rustin, M., 2013. After Neoliberalism: Analysing the Present. *Soundings* 53 (April) 8–22.

Hamdi, N., 2010a. *The Placemaker's Guide to Building Community.* Earthscan, London.

Hamdi, N., 2010b. Foreword, in: Lyons, M., *Building Back Better.* South Bank University: Practical Action Publishing, London.

Hamdi, N., 2004. *Small Change.* Earthscan, London.

Harcourt, W., Brooks, A., Escobar, A., Rocheleau, D., 2013. A Massey Muse, in: Featherstone, D., Painter, J. (Eds), *Spatial Politics: Essays for Doreen Massey.* Wiley-Blackwell, Chichester.

Harvey, D., 2010. *Social Justice and the City,* Revised Edition. University of Georgia Press, Athens.

Harvey, D., 2012. *Rebel Cities.* Verso, London.

Hatherley, O., 2014. The Problems with Favela Chic. *Disegno* 7 (18 December). Available at http://www.disegnodaily.com/article/the-problems-with-favela-chic (accessed 9 March 2016).

Hickey, S., Mohan, G., 2004. Relocating Participation Within a Radical Politics of Development: Insights from Political Practice, in: Hickey, S., Mohan, G. (Eds), *Participation: From Tyranny to Transformation.* Zed Books, London.

Hill, D., 2015. *Dark Matter and Trojan Horses: A Strategic Design Vocabulary.* Strelka Press, Moscow.

Jones, S., 2006. *Antonio Gramsci.* Routledge, London.

Kaplan, A., 1996. *The Development Practitioner's Handbook.* Pluto Press, London.

Kelly, U., 2004. Confrontations with Power: Moving Beyond 'The Tyranny of Safety' in Participation, in: Hickey, S., Mohan, G. (Eds), *Participation: From Tyranny to Transformation*. Zed Books, London.

Kipfer, S., 2008. How Lefebvre Urbanized Gramsci, in: Goodewardena, K., Kipfer, S., Milgrom, R., Schmid, C. (Eds), *Space, Difference, Everyday Life: Reading Henri Lefebvre*. Routledge, New York.

Koczanowicz, L., 2011. Beyond Dialogue and Antagonism: A Bakhtinian Perspective on the Controversy in Political Theory. *Theory and Society* 40(5), 553–566.

Krishnaswamy, R., 2002. The Criticism of Culture and the Culture of Criticism: At the Intersection of Postcolonialism and Globalization Theory. *Diacritics* 32, 106–126.

Laclau, E., 1990. *New Reflections on the Revolution of Our Time*. Verso, London.

Laclau, E., Mouffe, C., 2001. *Hegemony and Socialist Strategy*. Verso, London.

Latouche, S., 2010. Standards of Living, in: Sachs, W., *The Development Dictionary*. Zed Books, London.

Lefebvre, H., 1976. *The Survival of Capitalism*. Allison and Busby, London.

Lefebvre, H., 1996. *Writings on Cities*. Blackwell Publishing, Oxford.

Long, N., 1992. From Paradise Lost to Paradigm Regained?; The Case for an Actor-Oriented Sociology of Development, in: Long, N., Long, A. (Eds), *Battlefields of Knowledge*. Routledge, London.

Lummis, C.D., 1997. *Radical Democracy*. Cornell University Press, Ithaca, NY.

Lummis, C.D., 2010. Equality, in: Sachs, W., *The Development Dictionary*. Zed Books, London.

Marx, K., 1977. *Karl Marx: Selected Writings*. Oxford University Press, Oxford.

Massey, D., 1984. *Spatial Divisions of Labor: Social Structures and the Geography of Production*. Methuen, Basingstoke.

Massey, D., 1991. A Global Sense of Place. *Marxism Today* 38, 24–29.

Massey, D., 1993. Power-Geometry and a Progressive Sense of Place, in: Bird, J., Curtis, B., Putnam, T., Robertson, G., Tickner, L. (Eds), *Mapping the Futures: Local Cultures, Global Change*. Routledge, London.

Massey, D., 1994. *Space, Place and Gender*. Polity Press, Cambridge.

Massey, D., 1996. Double Articulation: A Place in the World, in: Bammer, A. (Ed.), *Displacements: Cultural Identities in Question*. Indiana University Press, Bloominton.

Massey, D., 1999. *Power-Geometries and the Politics of Space-Time*. Department of Geography, University of Heidelberg.

Massey, D., 2004. Geographies of Responsibility. *Geografiska Annaler Series B: Human Geography* 86, 5–18.

Massey, D., 2005. *For Space*. Sage Publications, London.

Massey, D., 2007. *World City*. Polity Press, Cambridge.

Massey, D., 2008. When Theory Meets Politics. *Antipode* 40, 492–497.

Massey, D., 2009. Concepts of Space and Power in Theory and in Political Practice. *Documents d'Anàlisi Geogràfica* 55, 15–26.

Mouffe, C., 1993. *The Return of the Political*. Verso, London.

Mouffe, C., 1999. Deliberative Democracy or Agonistic Pluralism? *Sociological Research* 66, 745–758.

Mouffe, C., 2013. *Agonistics: Thinking The World Politically*. Verso, London.

Parvin, Alastair, 2013. Architecture for the People by the People. TED talk (February). Available at https://www.ted.com/talks/alastair_parvin_architecture_for_the_people_by_the_people (accessed 9 March 2016).

Robbins, B., 2013. Subaltern Speak: Review of Postcolonial Theory and the Specter of Capital. *N+1* 18. Available at https://nplusonemag.com/issue-18/reviews/subaltern/ (accessed 9 March 2016).

Rustin, M., 2013. Spatial Relations and Human Relations, in: Featherstone, D., Painter, J. (Eds), *Spatial Politics: Essays for Doreen Massey.* Wiley-Blackwell, Chichester.

Sachs, W., 2010. One World, in: Sachs, W., *The Development Dictionary.* Zed Books, London.

Saldanha, A., 2013. Power Geometry as Philosophy of Space, in: Featherstone, D., Painter, J.Painter (Eds), *Spatial Politics: Essays for Doreen Massey.* Wiley-Blackwell, Chichester.

Schrijvers, J., 1993. *The Violence of Development.* International Books, Utrecht.

Sennett, R., 1970. *The Use of Disorder.* Penguin, Harmondsworth.

Sennett, R., 2003. *Respect, The Formation of Character in an Age of Inequality.* Penguin, London.

Smith, A.M., 1998. *Laclau and Mouffe: The Radical Democratic Imaginary.* Routledge, London.

Turnbull, S., 2002. *A New Way to Govern.* New Economics Foundation, London.

Turner, J.F., 1976. *Housing by People: Towards Autonomy in Building Environments.* Marion Boyars, London.

Turner, J.F., 1983. From Central Provider to Local Enablement. *Habitat International* 7, 207–210.

Westbury, M., 2015. *Creating Cities.* Niche Press, Melbourne, Australia.

Willis, J., 2013. Place and Politics, in: Featherstone, D., Painter, J. (Eds), *Spatial Politics: Essays for Doreen Massey.* Wiley-Blackwell, Chicago.

Yates, D., 1980. *The Ungovernable City.* MIT Press, Cambridge, MA.

4   Identity and practice

In previous chapters we have explored and compared both Turner's and Hamdi's development methodologies in the critical context of Lefebvre's and Massey's spatial theories. In this chapter we will look to build on this analysis and explore the implications of participatory development practice methodologies further by considering them in the context of a wider discussion of identity theory and question what can be learnt from the post-colonial context of global development.

The subject of identity is critical to our wider premise. The relationship of identity to space provides an opportunity to consider the spatial methodologies of innovative development practice against a broader interdisciplinary context of political and cultural theory. Ultimately this critical analysis will look to highlight the positive potential of Turner's and Hamdi's spatial methodologies and development practices as interconnected with the articulation of new cultural identities and social practices. The intention thus remains to reflectively learn from this analysis and begin to question the interdependence of identity and practice in connection with a re-imagining of Westernised space and architecture.

Our comparative and critical re-reading of key development discourses frames Turner and Hamdi in the context of the cultural diversity of the emergent spaces of the Global South. In doing so it provides a further framework with which to interpret and re-evaluate the theoretical and methodological value of Turner's and Hamdi's development practices as exemplars of post-structural spatial practice. The implications of the scale and complexity of this study negate the ability to offer a complete discourse on the historical evolution of development practice and its relation to Turner or Hamdi (Chang, 2002). As such, this chapter borrows from already well established critical comparisons in order to enable a clearer discussion of practical methodological implications of identity and the potential for positive development change.

Critical and detailed analysis of the history of development and colonialism is available in well established texts by Gilbert Rist (2006), and Majid Rahnema and Victoria Bawtree (1997).

Whilst engaging with the full breadth of discussion encompassed in post-colonial theory is not the intent of our study, the opportunity to critically appropriate themes and concepts from such theory provides moments of valuable reflective anlaysis. Thus, the open and discursive trajectory of the analysis in this chapter is proposed as an attempt to both provide a new open perspective on development practice methodologies, and to highlight key interdisciplinary points of comparison and reflections upon Western space.

Based upon these observations the relational interdependence of knowledge, power, authority, and identity is seen to offer a valuable point of comparison between the conception of identity as a product – a theoretical and ontological construction – and identity as practice – the valuing of cultural specificity and everyday life as found in the alternative development practices of Turner and Hamdi. In exploring such a critical analysis our study utilises Edward Said's pioneering work *Orientalism* (2003) in which academic discourse concerning the historical dialogue of East and West is critiqued as an institutional projection of identity and power. Said is recognised as providing much of the theoretical framework for subsequent post-colonial interrogations of the relationship between the occident and orient, West and East, us and them (Attridge et al., 1989; Eagleton, 2006; Rubin, 2003; Ruthven, 2003; Young, 2004, 1990).

Building on this theoretical discussion of the implications of Said's notion of political authority in development space as exploiting a 'flexible positional superiority' (Said, 2003, p. 7), this chapter contests that the development practices of Turner and Hamdi offer valuable insights in comparison to discourses of emerging counter-narratives of post-colonial identity and its economic and political contexts. Beginning with Turner, we will analyse the methodological shift pioneered in his models in Peru, observing the implications of his re-articulations of development from intervention to interaction, participation to partnership, housing to sustainable enterprise (Turner, 1983). Examples of practical observations and experiences drawn from key texts (1972a) help to elucidate this comparison and Turner's response to identify the implications of identity for his practice.

This analysis of identity is then broadened to explore the critical identity of development in relation to global political and economic homogenisation and hegemony (Massey, 2005, pp. 69–70), and the conception of 'under-development' as an identity. This observation is drawn into comparison against Hamdi's observations of the political and economic mis-appropriation of the promise of Turner's early development practice throughout the late twentieth century (Hamdi, 2010, pp. 2–9), before being explored against the implications of contemporary discussion of 'post-development' practice (Sachs, 2010a, p. xiii).

The critical articulation of the interdependence of social identity and participatory practice is then further explored in examples drawn from Hamdi's discourse, reinforcing the links between theoretical and methodological insights into identity. The various connections Hamdi makes to issues of vulnerability (2010, p. 53), dependency, ownership (2010, p. 180, 1989, pp. 23–24), and livelihood (2010, p. 190), are referenced in order to define comparisons between methodological tropes and their positive theoretical implications (Burnell, 2012, pp. 138–140).

Further critical comparison of both Turner's and Hamdi's participatory practice is then explored in the contestation of the negative implications exposed in participatory practice by Bill Cooke and Uma Kothari (2001) and the subsequent renewed positive potential by Samuel Hickey and Giles Mohan (2004a). This exploration of critical participation reflects Massey's critique of spatial conformity and inevitability in development assumptions and is here compared with Hamdi's renewed contestation of practice as a means to empower local sustainable enterprise and social relations of identity (Hamdi, 2004, pp. 83–85).

In drawing a critical conclusion to this chapter, the premise of 'identity as practice' is discussed in the context of a renewed critical consideration of the work of Turner and Hamdi. This comparison is articulated through a contextualisation of the deeper and previously un-connected theoretical discussions of both concrete practical realisations and concurrent theoretical articulations of identity and practice (Hamdi, 2010, p. 54, 2004, p. 26). The integration of this interdisciplinary critique of identity provides a valuable thematic layer of socio-political critique to our underlying methodological trajectory. It questions the contextual relations and implications of the political and historical processes of global inequality and capitalism, with the practical reality of identity, subjectivity, and equality confronted in alternative development practice.

Identity as a product

In order to critically contextualise Turner and Hamdi within a wider framework of interdisciplinary identity theory, we must first explore Edward Said's text *Orientalism* (2003). Said's ground-breaking analysis of academic discourse concerning 'the orient' is generally appreciated as one of the first critical applications of post-structuralist theory to historical documentations of the space and global politics of colonial empire (Robbins, 1993, pp. 58–59; Slemon, 2006, pp. 45–48). Its critique of the relationships between dichotomies of ontology and identity, discourse and action are widely acknowledged as having provided the theoretical foundation from which post-colonial and subaltern studies have evolved (Ashcroft et al., 2003, p. 85; Bhabha, 2004, p. 102; Clifford, 1980; Spivak, 1999, pp. 265, 270).

Said's documentation of this relationship between Western identity and a generic oriental *Other* (2003, p. 10) is interpreted in the context of our analysis as a clear precursor to the contestation of multiplicity in the post-structural geographies of theorists such as Doreen Massey (Ferguson and Gupta, 1992, p. 8; Massey, 2005, pp. 66–67). However it also intersects critically with notions of identity and values that have been instrumental in the criticism of Wolfgang Sachs (2010b), Homi Bhabha (2006, pp. 155–157, 1990; Moore-Gilbert, 2000), and Gayatri Spivak (1998).

In the context of these intersections the potential viability of our comparison with development practice becomes clear. Said is explicit in the description and analysis of orientalism's ideologies and its implications for the identity, freedom, and subjectivity of persons, governments, and organisations who interact with others and otherness (2003, pp. 210, 216, 1983, pp. 14, 29). Reflecting the same critical reservations as we highlighted from Hamdi in chapter three, Said notes the inability to engage in the physical or theoretical context that surrounds orientalism without implicating conflicts of class, race, religion, and socio-political history in the discussion (2003, pp. 10–12). No matter how well intended such actions might be Said is clear that because of 'Orientalism the Orient was not (and is not) a free subject of thought or action' (2003, p. 3).

Building upon key post-structural theoretical methodologies – specifically Foucault's analysis of the relationship between knowledge and power (Bhabha, 2004, pp. 103–106; Said, 2003, pp. 3, 14, 22–23) – Said's analysis frames orientalism as an expression of globalised socio-cultural inequalities, and subsequently as a political manifestation of identity and authority (2003, pp. 6, 10, 327). For Said, orientalism reflects the utilisation and manipulation of abstract academic discourse in order to represent and define the culture and identity of 'the other'. This analysis can be seen further reflected in the observations of Franz Fanon whose studies into the psychopathology of colonisation explored the human, social, and cultural consequences of de-colonisation (Abdilahi Bulhan, 1985; Fanon, 2001, pp. 169–175; Gordon, 1995).

Observing numerous historical manipulations of philological documentation Said observed this phenomenon as a politically motivated discourse that authorised, produced, and represented space and identity (Bhabha, 2004, pp. 118–199). Its product was an identity existing interdependently as both abstract discourse and concrete practical manifestations contributing to and reinforcing the politics and economics of global inequality (Said, 2003, pp. 6, 52).

This analysis allows Said to provide a critique of orientalism as being produced and maintained by the assumptions of Western authority (Jameson, 2001, p. 49; Said, 2003, pp. 86–87, 197, 2001, p. 73). Through extensive historical and geographical examples he demonstrates how the political implications of this exchange were used to justify economic, political, and geographical conquest and ideological hegemony (Lewis, 2007; Said, 2003, pp. 73, 92). Reinforcing Massey's advocacy of the multiplicity of space, this analysis can be critically re-appropriated within this chapter as a framework within which to read and contextualise the interdependent political context and evolution of development in the twentieth century (Hamdi, 2010, pp. 6–9).

Said's rich and detailed critique of orientalism as a structuralist and self-referential discipline ultimately exposes the conception that identity as a product can never be interpreted as an accurate representation of others (2003, p. 22). Instead the lack of critical self-reflection that defines orientalism as a theoretical abstraction allows such discourse to represent the reflective identity of the authors themselves. Thus, he observes orientalism as offering a critical representation of certain facets of the Westernised identity that remain hidden in the ideological cohesion and hegemony of capitalism (Eagleton, 2001, p. 26;

Said, 2003, p. 53, 2001, p. 78). It is in this theoretical inversion that Said explicates the contemporary implications of the moral authority derived from the definition of 'the other' (2003, p. xii). Considered in this way, the abstract and politically produced identity of the oriental *other* comes to represent little more than a manifestation of all the things that the West despised and feared in and of itself:

> Along with all other peoples variously designated as backward, degenerate, uncivilised, and retarded, the Orientals were viewed in a framework constructed out of biological determinism and moral-political admonishment. The Oriental was linked thus to elements in Western society (delinquents, the insane, women, the poor) having in common an identity best described as lamentably alien. Orientals were rarely seen or looked at; they were seen through, analysed or confined or – as the colonial powers openly coveted their territory – taken over.
>
> (Said, 2003, p. 207)

This form of self-denial in the negative identification of the oriental other (Said, 2003, p. 204) thrives on the ability to associate the products of its own moral inequalities[1] with the fear and danger embedded within a distant, shadowy, and indefinable other (Guha and Spivak, 1989; Young, 2004). This contestation becomes intriguing when global development is critiqued as relying upon similar universal identities of 'under-development' (Lummis, 2010, p. 51), and 'catching up with the West' (Massey, 2005, p. 68). Within this observation is a recognition of the inequality of Western moral authority as a product of geo-political history and an explicit reluctance to engage with or confront the reality of spatial multiplicity as explored by Massey (2005, p. 82, 1994, p. 22).

Why is this analysis of colonialism and the politics of identity important to our aspirational re-imagining of space and architecture? Said presents orientalism and more broadly identity as a means to isolate and disassociate Western ideology from the negative identity of 'the other', and this ideological disassociation from others can begin to be perceived as interconnected with concepts of development and culture, whether it be in the Global South or more conventional Westernised spaces. Building on this analysis we can now begin to contest whether development can ever be more than a representation of the West, and thus always some form of projection upon its recipients. And conversely, what does this realisation of the self-reflective representation of authorship of that development (and architecture more generally) imply for any critical re-imagining of space and architecture in the interconnected and globalised multiplicity of the twenty-first century?

Here it is important to note various counter-arguments and challenging re-contextualisations of the broad implications and suppositions Said draws in his analysis. Such critiques are notable for their explication of both the positive and negative aspects of the moral subjectivity placed upon the authors of political history. Ibn Warraq contentiously notes the implications of what he describes as Said's oversimplification of the orient/occident relationship into a strict dichotomy (2007, pp. 40–43). Whilst this in itself appears to overlook the clear intentions to not engage in any such dichotomy as outlined in Said's rebuttal to such criticisms,[2] our analysis finds sympathy towards Warraq's concern for the multiplicity of subaltern identities that are subsumed under the academic scale within which Said's discourse theoretical inversion of oriental discourse is implicated (2007, pp. 23–24).[3]

Yet Said's critique of orientalist identity as a product and negative self-reflection of the West might still be used provocatively in direct comparison to the more practical applications and implications of identity and the dynamics of global power relations of inequality and development (Esteva et al., 2013; Lummis, 2010). Simply put, the negative projection from the occident upon the orient provides a continuing physical and theoretical manifestation of authority that persists in today's global politics and explicitly affects the discourse and practice of international development.[4]

Orientalism's un-critical use and manipulation of dualistic and negative otherness generates a lasting and endemic sense of 'us and them' that remains deeply engrained in contemporary socio-political questions (Said, 2003, p. 327). It is a singular vision of history, development, identity, and democratic moral superiority that continues to be projected using discourse as a tool of hegemonic negation and subjugation of the other. This is similar to Massey's contemporary contestations of space as the medium for political and relational space (2005, p. 87), which concurrently denies the multiplicity and directionality of power whilst enacting the suppression of cultures that cannot not be made to conform:

> It is hegemony, or rather the result of cultural hegemony at work, that gives Orientalism the durability and the strength I have been speaking about so far. [...] In a quite constant way, Orientalism depends for its strategy on this flexible positional superiority, which puts the Westerner in a whole series of possible relationships with the Orient without ever losing him the relative upper-hand.
>
> (Said, 2003, p. 7)

This notion of a 'flexible positional superiority' is the clearest explanation of how abstract discourses can be shown to continually produce and reproduce identities that are constructed around simplifications of right and wrong (Spivak, 2004, p. 532), us and them, here and there, developed and developing (Lummis, 2010, p. 51; Sachs, 2010b, p. 5). The effects of such sustained and proliferated discourses of superiority are keenly felt in the expressions of identity that they historically produced and that continue to be contested in contemporary development as an ideology (see 'Common sense, identity, and culture', in Hall et al., 2013). In responding to shifting historical contexts a politically constructed orientalism validated a moral and authoritative suppression of otherness that again warrants comparison to the history of development practice and the concept of development and its corollary under-development (Escobar, 2010, p. 145; Fabian, 2002, p. 143; Roy, 2011, p. 224).

Echoing Said's critique, our wider research suggests that if development discourse can be perceived as a metaphorical mirror of the Western social conscience, it reveals a representation of a systematically constructed global capitalism and economic inequality. It reflects and highlights a construction of negative identity that is reliant upon a produced identity of universal Western values that invalidates the multiplicity of identities and practices that prosper in the alterity of other, different, and alternative spaces and cultures. Thus whilst Said's historical and theoretical implications of colonial empire and the production of identity allow a philological critique of the subjectivity of history, it also offers the opportunity for comparison with the production and manipulation of identity suggested in the contentious history of development practice and the residual contemporary continuity of such negative articulation of otherness and difference (Said, 2003, p. 327).

This brief exploration of Said's critique of orientalism provides valuable critical frameworks with which to compare both positive examples and negative critical observations of the appropriation and contestation of identity in development practice. Here, questions of authority, control, and freedom that Turner raised can be compared with post-colonialism's critique of 'negative identification of Others' (Bhabha, 2004, p. 249), and the underlying 'flexible positional superiority' (Said, 2003, p. 7) that pervades orientalism. Placed in a critical comparison with Turner's observations and critique of development practices, the same projection of negative identity and flexible authority might be observed as pervading the evolution of global development policy and practice over the fifty years since his original observations.

Furthermore, what does the notion of 'flexible positional superiority' begin to imply to the conventional articulation of architecture and space in Westernised contexts? If identity and development practice are linked, then architecture is implicated in a far more complex relationship of cultural identity than is conventionally recognised. Echoing our observations about Marx in chapter two: we produce space (architecture, cities), and the space reflectively produces us. Thus, to re-imagine space we must acknowledge our complacency about identities of otherness and alterity that are so prevalent, rich, and positively challenging to development practice.

Turner and identity

John Turner's methodologies of participatory practice can be considered as a key demarcation of the origin of criticality in post-colonial development space. His work marks a watershed moment in the shift from colonialism to development in the 1950s and 60s (Harris, 2003, pp. 245–248; Mathey, 1991; Peattie and Doebele, 1973; Rodell et al., 1983). This historical change is notable for corresponding with Turner's research and presentation of the seminal paper 'Uncontrolled Urban Settlements' (1968a) at a United Nations seminar in 1965. His subsequent research at MIT and involvement in the UN development policy frameworks are similarly recognisable as a pivotal point in the theoretical discourse of development practice, and Turner's research is perhaps the most influential contribution to setting in motion governmental 'sites and services' housing programmes (Peattie, 1982).

Both Turner's socio-economic observations and his practical realisations of alternative development were explicitly based upon the importance and value of choice and specifically, the freedom to choose to build your own home (Turner, 1976, pp. 11, 54, 61, 153–154). As observed in chapter one, Turner not only provides explicit and evidenced critique of the socio-economic implausibility of hierarchical development intervention (1976, pp. 14, 42, 1974, 1972b, p. 169, 1968b), but crucially also a positive alternative methodology based upon the critical political frameworks of autonomy and heteronomy (1976, pp. 17–19), and the implications of mismatches of identities and values:

> Quantitative methods cannot describe the relationships between things, people and nature – which is just where experience and human values lie. [...] Only by standing Lord Kelvin's dictum on its head can one make sense of it: nothing of real value is measurable.
>
> (Turner, 1976, p. 64)

This advocacy of progressive development has been conventionally valued for its remarkably clear and evidenced demonstrations of both the social and economic value of self-build and user-defined housing (Turner, 1976, pp. 43–48, 1972b, pp. 163–165, 1963, pp. 363, 381, 381). Whilst not wanting to overlook criticisms raised by Burgess (1978, 1977, p. 1117) and retrospectively by Peattie (1982), simply put, progressive development practice advocates housing and communities organised, built, and managed by the inhabitants of informal settlements themselves (Turner, 1972b, pp. 154–160).

Crucially this approach stresses the necessity of grass-roots, participatory, and 'bottom up' approaches, coupled with strategic and political advocacy of the democratisation of planning legality in order to support the social production of space and development via collective and self-management (Fichter and Turner, 1972, p. 127). This discourse thus reflects a re-evaluation of informality as a positive alternative to Western identity and development models of top-down institutional and hierarchical policies (Turner, 1983, p. 208). Here Said's critical analysis of abstract and negative identification of otherness and the interdependent reproduction of flexible positional superiority are crucial. This analysis of post-colonial space frames a critical analysis of the historical and political contexts that are implicated by development's global and local relationships. In framing the theoretical connection of identity production and inequality in this way, our analysis in this chapter posits development as interdependent with an adaptive framework of geometries of power and values (Massey, 2005, pp. 84, 103).

For Turner this experience is clearly articulated in his discussion of a project for a new school in Tiabaya near Lima in Peru (1972a, p. 125). The implications of such geometries of power were uncovered in the problems he faced in his aspirations for this school. As you would expect, Turner proposed what he believed to be a contextually, materially, and economically efficient design for the school based on the inherent appropriateness of the vernacular building style, local materials, and well-intentioned thoughtful design.

Retrospectively however, Turner notes that his approach in communicating the design to the village council ended up being an 'onslaught of economic and design logic', before ruefully writing: 'our own enthusiasm was not audibly echoed or even shared by the council members' (1972a, p. 126). Upon returning to the village to see work progressing, Turner noted that the designs had been changed by the council in his absence and were now attempting to use concrete and steel whilst keeping just his overall layout (1972a, p. 127). The implications of changes of material would mean that the project would not be possible on budget and was ultimately doomed to failure. Crucially, instead of decrying such happenings as the naivety of other people Turner turned his frustration inwards and was self-critical and reflective:

> The disaster which would have overtaken the well-intentioned Tiabaya school project [...] would have been the direct result of power to impose decisions from above which must come from below if good use is to be made of local resources. [...] We, the authorities, overpowered the Tiabaya School committee with words, and though more respectful of our Miraflores clients' felt architectural needs, we overwhelmed them with our political power.
>
> (Turner, 1972a, p. 133)

Turner ultimately recognised the assumed authority used to impose a design upon the community that did not reflect their aspirations for a modernity that the school represented for the village community (1972a, p. 134). In being deaf to these local aspirations Turner saw in his own practice the implications of development as a projection of identities and assumptions. The best intentions of development cannot be reconciled through professional confrontation any more than they can overcome the manifest differences of multiple identities through brute force of design and values as an imposition upon other people (Turner, 1976, p. 23, 1972a, p. 145). In place of assumptions and authority Turner recognised what was needed was mutuality and humility.

In response to these experiences Turner was to explore an approach to development as a facilitation of autonomy, choice, and heteronomy (1976, p. 23, 1968b, p. 355) that built upon participatory engagement, political and economic education, empowerment and advocacy, and more fundamentally the contestation of assumed roles, identities, and values. In his important observations of informal housing development Turner would note the mismatches and prescriptions of value implicated within conventional top-down housing models:

> If the usefulness of housing for its principal users, the occupiers, is independently variable from the material standards of the goods and services provided as the case studies and other sources show, then conventional measures of housing value can be grossly misleading. As long as it is erroneously assumed that a house of materially higher standards is necessarily a better house, then housing problems will be mis-stated.
>
> (Turner, 1976, p. 60)

Various examples of the success of such projects can be found in Turner's self-help housing in Huascara in Lima, Peru (1972a, p. 141), in the Brazilian Algado housing system (1974), and in informal barrios that Turner observed in Caracas (1968b, p. 361), each representing a version of self-management and grass-roots social practices of material necessity. Yet these observations hint at the problematic implications that such alternative development practices suggest for socio-cultural relations and global political inequality of identity and value. Here there remains the implication of a freedom and necessity to build not only your own home, but also your own identity (Turner, 1976, p. 23).

The disjunction between produced identities based upon abstract assumptions,[5] and practised identities of material reality reveals a connection to the political authority and control in development discourse and international policy. This crisis of identity in development that can be observed pervading and persisting in mainstream discourse (Rahnema, 2010a) is searingly and darkly reiterated by Turner's humble and stark observations of ideological development and housing in informal settlements (1974, p. 61). He describes the culmination of misplaced good intentions, political and economic hegemony, and the underlying premise of a singular, universal purpose and identity pertaining to development:

> The vast majority of officials and professionals keep recommending the destruction of people's homes in order to solve these same people's 'housing problems' by providing them with alternatives either they or society cannot afford. In a world of grossly maldistributed resources and injustice, this is a huge, but very black joke. Such stupidities are inevitable as long as those who perpetuate them have confused their values and lost their common sense of life's wholeness.
>
> (Turner, 1976, p. 61)

These dark observations of misplaced values are contrasted by the positive potential Turner highlights in examples of development identities that are inherently counter-intuitive to conventional ideologies of development. In contrast to formal legal and planning infrastructures such examples highlight the support, empowerment, and integration of informal settlements as positive and alternative realisations of material necessity and grass-roots sustainable enterprise (Hamdi, 2010, p. 3; Turner, 1968a, pp. 121, 127). Turner explores these observations variously in his discourse, but an exemplar of the efficiency of informal housing and identities can be found in his analysis of Mama Elena's low-income communal household (1976, p. 90).

The example of Elena describes the experience of a family whose frequent forceful eviction from unaffordable government tenement buildings by state agencies and police, led to the communal creation of 'provisional shack' dwellings that existed in a highly convenient yet formally illegal situation close to schools and work (Turner, 1976, p. 91). This illegal and unorthodox system afforded a degree of flexibility and adaptability to circumstances that was never afforded by previous tenement occupancies. It thrived on variables of social and spatial relations that were based in material and economic realities that were contradictory to external agencies and government perceptions of what development looks like (Turner, 1974, p. 96). Turner's involvement was not as facilitator or developer, but merely as political advocate for this and other projects as examples of materially and socially sustainable practices and identities.

For Turner, effective housing and development was a product of 'what housing does for people' and not 'how it looks' (1974, p. 97). He realised that a house can only act and succeed as a home if people's housing needs are stated in terms of material and social priorities of *access, shelter,* and *tenure.* These three variables might be considered largely universal and independent of formal or informal housing and development (Turner, 1976, pp. 51, 63, 153–154).

Here we might suggest an opportunity for renewed critical awareness of examples of alternative housing in the UK, and perhaps most notably the plotlander movements that were exquisitely documented by Colin Ward. Written with frequent collaborator Dennis Hardy, *Plotlanders* (1972) provides a documentation of the emergence of informal seaside housing that began in 1907 at Pagham Beach, Selsey in Sussex. This example of informal seaside homebuilders provides a vehicle for Ward to discuss the implications of plot-landers as a social movement (and a positive example of anarchist space), before documenting the slow choking of this way of building and living by post-Second World War institutional planning policy. Ward notes:

> The word 'plotlands' is used by town planners as a shorthand description for those areas where, in the first forty years of this century, land was divided into small plots and sold, often in unorthodox ways, to people wanting to build their holiday home, country retreat or would be small-holding. Sometimes they simply squatted and eventually gained title through 'adverse possession', the legal phrase for squatter's rights.
>
> (Ward and Hardy, 1972, p. 63)

The emergence and continued (if somewhat compromised) existence of such informal housing and development seems so devoid from contemporary life in the UK (and wider Westernised world) as to seem incongruous and divorced from modern housing problems (Mann, 2003, p. 111). However this short-sighted perspective is highly skewed by current cultural hegemony surrounding the commodification of housing by capitalist politics and neoliberal economics (Harvey, 2015, pp. 15–24).

It might seem that the eradication of informal housing in the UK is a sign of development that the rest of the world is merely waiting to catch up to; a premise that Doreen Massey and Wolfgang Sachs et al. would strongly disagree with (Massey, 2005, p. 4; Sachs, 2010b). Yet with the inexorable growth of informal housing in the Global South coupled with endlessly increasing global (and local) inequality the conditions of informality and the history of plotlanders perhaps are increasingly applicable to the social conditions of Western space. Consequently we might find valuable insights by engaging with and valuing alternative ways of living (Ward and Hardy, 1972, pp. 67–68), and thus with reference to informal housing in the UK we can once again use Ward's detailed insights on the origins and potential of plotlanders:

> The word 'plotlands' evokes a landscape of a gridiron of grassy tracks, sparsely filled with bungalows made from army huts, railway carriages, shanties, sheds, shacks and chalets, slowly evolving into ordinary suburban development.
>
> (Ward and Hardy, 1972, p. 63)

Parallel to Ward's documentation of alternative and anarchistic housing in the UK, Turner offers similarly explicit references to the global political implications of the inequalities and implausibility of development as a singular homogenous identity. In critiquing the pinnacle representations of development assumed to be Westernised space and nations Turner suggests an inversion of the political, social, and economic validity of neoliberal capitalism that resonates with our underlying critiques of Westernised spatial practices: 'This apparent paradox, created by false values and confused language, is a very common one, especially in the majority of low-income countries as well as, and perhaps increasingly, in countries like Britain' (Turner, 1976, p. 52).

In this reflective critique of Western values and identity Turner (and his contemporaries like Ward) hint at the same reversal of negative and presumed production of identity that pervaded Said's study of orientalism, and a profound geo-political inversion based upon Turner's critical observations and experience of conventional development methodologies. Favela, slum, and informal housing is all too easily dismissed as negative, degenerate, broken, and other to accepted normality. And similarly Ward observed the same social, economic, and cultural stigma and isolation in his analysis of the origins and slow choking of plotlander housing in the UK. The political implications of such a critique perhaps suggest why the same critique of identity still pervades contemporary development discourse (Sennett, 2004, p. 91), though is almost unheard of in connection with contemporary Westernised space. Some fifty years after Turner first critiqued the produced identity that accompanies development, which he calls a 'mirage' (1976, p. 15), we find Hamdi observing: 'the concept of bringing civilisation (development?) and promoting progress being a crusade (for some) resonates still with some of the ambitions, if not policies, which underpin the politics of aid under the guise of development' (Hamdi, 2010, p. 1).

'Under-developed' is an identity

The political question of identity in global development remained implicit throughout Turner's early participatory methodologies, but has since become pervasive in both practical and theoretical development discourse (Esteva et al., 2013, p. 3). Yet it is not until the late twentieth century context of expanding post-colonial and subaltern studies that development discourse began a renewed political and economic critique in reaction to this issue of development as an identity. This strand of discourse came to be identified as 'post-development' (Sachs, 2010a). Sachs' (et al) articulates a theoretical (and impassioned) critique of the apparent misconceived neutrality and passive subordination that pervades twentieth-century development discourse, policy, and practice:

> According to them, the 'backward' or 'poor' countries were in that condition due to past lootings in the process of colonisation and the continued raping by capitalist exploitation at the national and the international level: underdevelopment was the creation of development.
>
> (Esteva et al., 2013, p. 5)

Advocates of post-development contest that the post-colonial contexts of development in the Global South are suffused with political and economic projections of continued negative colonial identities that are implicated in all attempts at development practice. Thus, conventional development as a projection from one culture upon another (regardless of good intentions) is considered by many proponents of post-development as merely the translation of colonial oppression to a similar yet more subtle and duplicitous control in the concept of global development (Esteva et al., 2013, p. 54; Illich, 2010, pp. 99–102; Latouche, 2010, p. 279; Rahnema and Bawtree, 1997).

The most notable aspect of this critique surrounds the analysis of the political articulation and projection of negative identity that pervades the political, economic, and semantic identity of what it means to be 'developed' (Esteva et al., 2013, p. 3). The relationship of identity and the projection of Western assumptions of development is thus integral to our book's wider premise of re-imaging space and architecture as engagements with difference, multiplicity, and otherness.

The practical spatial reality and critical origins of post-development discourse is generally construed as originating in the inaugural address of President

Harry S. Truman to the United States and the rest of the world in 1949 (1967). Post-development protagonists cite this speech as a point of origin for the production of the negative political terms and identities of 'developed' and the corollary 'under-developed'. The structural dualism of these terms and identities is analysed as having fundamental ramifications for global politics (Massey, 2005, p. 68). The desire and goal to help the world attain the vision and idea of development that the Western world represented is perceived as enshrining the singular identity of development with the ideological cohesion of neoliberal values and capitalist mechanisms. Whether unintentionally or not, subsequent identities of development were once more articulated in a political monologue of what development meant:

> Underdevelopment began, then, on 20 January 1949. On that day a billion people became underdeveloped. In a real sense they stopped being what they were, in all their diversity, and were transmogrified into an inverted mirror of others' reality: a mirror that belittles them and sends them off to the end of the queue, a mirror that defines their identity, which is that of a heterogenous and diverse majority, simply in the terms of a homogenising and narrow minority.
>
> (Esteva et al., 2013, p. 2)

Academic and political responses to such post-development discourse have critiqued the oversimplification that is presupposed by the assumption that all development is imposed from the West as a spatial and political ideology (Edelman, 1999, pp. 10–15). This critique of such reductive dualism is perhaps validated by Marc Edelman's well documented argument that a large proportion of development can be observed arising from within the developing world itself, with Ray Kiely observing that '[t]he post-development idea is thus part of a long history within the development discourse' (1999, p. 30).

In another critique Kiely (1994) explores suggestions of similarities of post-development with neoliberalism in their rejection of top-down, centralised approaches. Such suggestions contend that radical decentralised governance might unintentionally be ignoring the potential of large scale strategic projects to assist impoverished people, instead placing them in economies of absence completely responsible for their own prosperity. The simple argument being that as we played a large contributing role in creating development disparity we should not abandon others to attempt to fix our mistakes.

In spite of the implications of these reservations, in the context of our comparisons of Turner's and Hamdi's local and strategic approaches to development practice, post-development theory still offers valuable insights into the contemporary contestation of post-colonial identity. Resonating with Massey's critique of the academic and political taming of space towards temporal convening and inevitability (2005, p. 70), the premise remains that the intrinsic assumptions of the hierarchical models and ideologies observed in conventional development are interdependent with the political and economic ideologies that accompanied efforts of post-colonial superpowers – a premise that is remarkably similar to observations made on the contemporary consequences of orientalism (Said, 2003, p. 327). The implications of perceiving development as an ideological construct are articulated in this book in order to reinforce the necessity of re-imaging space and architecture as a means to practically engage and confront the politics and material reality of identity, and to start by learning from the spatial observations, practices, and methodologies highlighted in Turner's and Hamdi's discourse.

In the context of these critical comparisons, the implications of inter-connected spatial identity and practice can again be perceived as playing out through the frameworks of imposing identities of those deemed as under-developed against a universal model of development framed around Western identities, concepts, and values. As noted extensively by Marianne Grone-meyer (2010), the consequences of this conflation of ideological development with local social values and identities are significant. Here Hamdi reiterates this issue:

> The phase of centralised planning and the public provision of everything including sites and service paralleled, more or less, the 1950s era of modernisation. When the ideals of modernisation were exported to the developing world they were done so on a simple assumption. If you want to be developed and 'modern' (like us), then do as we do, conform to how we do it in technology and style, use the standards and goals we set our-selves, adopt our vision of a better world and, in time, with a bit of luck and a lot of help (from us) you will achieve modernity!
>
> (Hamdi, 2010, p. 10)

In spite of the economic and political success of Turner's practices, the sub-sequent geo-political history of development as an idea and identity have become conflated with various capitalist and neoliberal policies emerging from various agents of change, for example the World Bank, corporate and institutional NGOs, and state-controlled aid programmes (Hamdi, 2010, pp. 8–12). The variety of projects based upon Western political and capitalistic values and development identities can be seen repeated again and again throughout later twentieth-century and contemporary development discourse (Burgess et al., 1997b, pp. 111–116; Illich, 2010, p. 106; Turner, 1976, pp. 13–18).

As noted by both Turner and Hamdi (amongst others), the identity and values that accompanied development can be perceived as interdependent with global capitalist economic policies. This relationship has played out through various implementations throughout the late twentieth and early twenty-first centuries (Eversole, 2012, p. 32). Here Hamdi is notable for pro-viding valuable introspection into the historical evolution of development practice, describing the shifting focus and emphasis of theoretical discourse and international policy (2010, p. 8). The success of development came to be judged based on criteria that relate directly to the Western model of what being developed means, through various shifting uses of GDP and global indexes of economic criteria. The dislocation from local values and identities of practice is here noted starkly by Lummis:

> The essence of economic development equality is contained in the phrase 'catching up' or 'narrowing the gap'. [...] The accusation of injustice cannot traditionally be made against inequalities between systems, but only within a system. The fact that the idea is intelligible today is evidence of the degree to which we accept that the world has been organised into a single economic system. [...] The idea that now the world economy has become capitalist it can generate quality through its own 'development' is remarkable.
>
> (Lummis, 2010, p. 145)

As observed in the discourses of Hamdi (2010, pp. 3–9) and Burgess et al. (1997a), the historical evolution of development ideology is frequently critiqued as a neo-colonial, capitalist, and hegemonic projection upon the developing world, and as an experiment of almost fundamentalist ideological conviction of the inevitability and infallibility of global capitalist modes of production (Esteva and Prakash, 1998, pp. 93–94, 152–153; Massey, 2005, p. 65, 2004, p. 8). Thus at no point in Hamdi's critique of the evolving ideologies of policy, economics, and politics of development can he identify an engagement with the vast range of differing identities that exist across the Global South. Instead global contexts of multiplicity and specificity are readily subsumed by abstract policy terminology and the assumed inevitability of neoliberal development and capitalist growth.

As such, under-development and its corollary of subsequent policy conflation of development as conforming and 'catching up to the west' (Massey, 2005, p. 68) are comparable to the various observations of the projected positive identity of development existing as a *mirage, guise,* and *fantasy* of either Western or capitalist visions of modernity and development observed by Turner (1976, pp. 14–15). However, more pronounced and immediate implications are surely felt in the negative corollary identity of under-development as envy and inferiority that remains largely uncontested and destructive within conventional cultural preconceptions and practices of development (Hamdi, 2004, p. 44).

Development identities and equality

Post-development discourse offers a remarkably similar articulation to the political observations drawn by Massey in the notion of 'temporal convening' of space and development, and an assumed inevitability of capitalist hegemonic development (2005, pp. 70, 82). In this conflation of time, development, and equality (Lummis, 2010) there is notably a similar dismissal of the specificity of space and context, and subsequently the identity observed in the historical abstractions found in orientalism. The articulation of what it means to be under-developed abruptly casts two-thirds of the planet's population with a single identity, overlooking the cultural uniqueness and multiplicity that exists within those who are 'other' (Esteva et al., 2013, p. 3; Lummis, 1997, p. 67). Building on this abstract universalism of identity Lummis critically connects the implications of catching up with the West with ideological constructions of economic inequality:

> Placing all the world under a single yardstick, so that all forms of community life but one are disvalued as underdeveloped, unequal and wretched, has made us sociologically blind. [...] How and when a people prospers depends on what it hopes, and prosperity becomes a strictly economic term only when we abandon or destroy all hopes but the economic one.
> (Lummis, 2010, p. 51)

Much like Said's observations of orientalism as an abstraction and negation of identities in support of economic and colonial conquest (Lewis, 2007, p. 778; Said, 2003, p. 216), Lummis proposes that the transcription of Western hegemonic values as economic ideologies of development produced a quantifiable scale with which to distinguish between developed and under-developed, North and South, us and them. Such an economic measure is devised with a scope and scale that applies irrespective of the global multiplicity of difference and otherness.[6] Thus, our critical comparison of the production of identity with Turner's observations of the use-value of development are valuable here in recognising that 'No one denies the universal need for homes any more than the importance of learning or keeping in good health. But many have come to identify the ends with the ways and means that turn them into products' (Turner, 1976, p. 12).

Such comparisons with post-development discourse offer a crucial challenge to the validity of the theoretical discourse and policy that frames international development as a global practice (Turner, 1983, p. 209). As explored in our earlier critical comparisons, the question of the values, meaning, and identity politics that accompany development are exemplified in the alternative spatial practices articulated by both Turner and Hamdi. Yet the significance of conflating a moral authority and geographical universalism of development with such an apparently universally acceptable concept of equality remains hidden within the complex rhetoric and semantic hegemony of development policy (Lummis, 1997, pp. 62–64). Thus for some the continued existence and political utilisation of development itself became a cause for protest and highly charged academic contestation, as exemplified in the claims of Majid Rahnema, Arturo Escobar, and, in this example, Ashis Nandy:

> The underlying myth of development, that it will remove poverty forever from all corners of the world, now lies shattered. It is surprising that so many people believed it for so many years with such admirable innocence. For even societies that have witnessed unprecedented

prosperity during the last five decades, such as the United States of America, have not been able to exile poverty or destitution from within their borders.

(Nandy, 2002, p. 108)

This shift in the critique of equality from distant alterity of the under-developed identity of otherness, to an introspective within the supposed sanctity of developed nation states is critical. Much like Turner's earlier reflection of housing values and development identities back upon Britain (1976, p. 52), such a contestation of equality suggests that Western development might no longer be represented as a universal aspiration (Sachs, 2010c, p. 124). Instead, what becomes crucial for the identity of development and equality are the processes, products, and implications of inequality experienced across unequal geometries of economic power, and gaps of inequality that exist throughout both supposedly developed and developing countries (Massey, 2005, pp. 82–84, 2004, pp. 11–16).

In essence, the identification of inequality cannot be limited to national boundaries and must instead confront the universal struggles of unequal power geometries created by the economic systems of employed in the name of neoliberal development (Bauman, 2000, pp. 15, 70; Harvey, 1996; Massey, 2004, p. 9, 1994, p. 22; Young, 2011). Thus the interdependence of identity and practice are as integral to critical questioning of Westernised space as they are to development in the Global South.

Turner explicitly outlined the implausibility of this picture of development as a means to catch up with the West in the 1970s, and showed through his observations of economic and cultural oppression that apparently well-intentioned development could be a fundamentally damaging and detrimental approach (1976, pp. 39–42, 1967, pp. 167–170).

The neoliberal capitalist principles of trickle-down development and the premise of capitalist hegemonic inequality being a viable model from which to generate a globally distributed financial equality (Esteva et al., 2013, pp. 23, 68) are now being actively re-contextualised in light of post-colonial and post-development academic discourse (Burgess et al., 1997c; Esteva and Prakash, 1998, pp. 22–26). Quoting Lummis here at length, we can observe a valuable comparison to the original observations and advocacy of Turner, noting how his advocacy of politically alternative progressive developments (1986, p. 20, 1974, p. 8, 1968a, p. 127) based upon material and social use-value compares favourably with many of the post-development critiques of universal values and equality:

> Development equality – catching up with the rich through economic activity – is thus a notion that goes against both common sense and economic science; it is a physical impossibility (assuming the earth is only one planet we have) and a logical contradiction. At the same time it operates, in fact, to establish a new form of inequality. Placing the world

under a single standard of measurement, it destroys the possibility of what might be called 'the effective equality of incommensurables'. For if it could be recognised that different cultures really have their own standards of value, which cannot be subsumed into one another or rank-ordered on some supra-cultural scale, it would make sense to give each equal respect and equal choice. The contrary notion, and the prevailing one today, that all the world's cultures can be measured against a single 'standard of living' measure (which implies standardisation of all living) renders all those cultures commensurable, and hence unequal. It dispossesses the world's peoples of their own indigenous notions of prosperity.

(Lummis, 2010, p. 50)

As with Hamdi's practical discourse, here the cultural specificity of place and identity can be observed as interconnected with a multiplicity of ways to perceive value and success in development (2010, pp. 44–45). Thus, issues of equality can begin to be interpreted as interdependent with the positive potential of otherness as an identity of development, and in direct contrast to the universal capitalist image and ascription of exchange value (Esteva and Prakash, 1998, p. 126; Turner, 1976, pp. 64–72). This analysis reframes the premise of development through outside intervention as not merely a continuation of a global economic equality within a capitalist context, but also as a highly politically motivated coercion of freedom, identity, and prosperity (Parry, 2006, p. 44; Williams, 2004).

Based upon this critique, the cohesion of capitalism as a vehicle to realise equality through development is implausible (Lummis, 2010, p. 51; Rahnema, 2010a, p. 183), but perhaps more importantly it highlights and reveals an ideological mis-direction that perpetuates the same reflective identity construction of negative difference and otherness that Said observed in orientalism. This manipulation and projection of an overtly over-simplistic distinction between those who have and those who have not underpins the ideological premise of development and under-development (Grenell, 1972, pp. 108–109; Illich, 2010; Turner, 1972b, p. 152). The perception of equality in a dichotomy with development leads to the simplification that supposes the rejection of diversity for the sake of perceived universal utopian ideals. Thus as Esteva contends:

But for two-thirds of the people on earth, this positive meaning of the word 'development' – profoundly rooted after two centuries of its social construction – is a reminder of what they are not. It is a reminder of an undesirable, undignified condition. To escape from it, they need to be enslaved to others' experiences and dreams.

(Esteva, 2010, p. 6)

This conception of identity as the means to development equality thus ignores the underlying truth that Western existence is built upon the inequality that we propose to reduce (Rahnema, 2010a, p. 183). Development's utilisation of a universal articulation of concepts of equality, needs, poverty, and growth are each used to maintain an ideological 'flexible positional superiority' so that the structural conditions of capitalist global power-geometries are proliferated as identity (Gronemeyer, 2010, p. 55). Thus Lummis observes the significance of post-development theory in articulating this disjunction between ideological development equality and its systematic construction of negative identity as a projection of abstract value judgements:

> Equality as justice is a value statement concerning how people ought to be treated; it refers to relations between persons. Equality as sameness, however, is an allegation of fact; it postulates common characteristics in people. A value statement may be derived from it. However if equality as sameness is asserted as a value, it may turn out to allege not a fact that is, but a fact that ought to be created. When this notion becomes attached to power, the consequences can be frightening.
>
> (Lummis, 2010, p. 38)

The implication of equality as a value statement based upon abstract and universal rights is observed as both theoretically and practically powerful and dangerous. Universal notions of how people ought to be treated such as the declaration of human rights to democracy and freedom from persecution and so on are incontestable in their global value. However a post-development analysis of equality as a value of sameness suggests that development had been conflated with an identity based upon an implausibility of universal equality that exists at the heart of capitalist ideology (Lefebvre, 2009, pp. 105–111, 205–206).

The critical connections and comparisons offered so far in this chapter have sought to contextualise the socio-political contestation of identity as a product and project of catching up with the West. This analysis reflects the potential of development to be articulated as a product that continues to inherently proliferate inequality through inevitability and universal values. In contrast, Turner's and Hamdi's observations and advocacies of distributed and grass-roots development practices are exemplary in the critical self-awareness of the positive potential of alternative, informal spaces and identities. It is here that both Turner and Hamdi are notable as offering something different in their advocacy of autonomy and heteronomy of progressive and sustainable development. The political implications of development as democratic and participatory practice are thus made interdependent with a grass-roots socio-economic inversion of neoliberal capitalist policies. Such acts articulate an inversion of the application of identity as a *product* of development, instead proposing identity as interdependent and contingent upon the *practice* and *process* of development, and thus potentially ripe with positive potential for engaging in the pluralism and multiplicity of post-structural space.

Hamdi and identity

Both of Hamdi's key texts, *Small Change* (2004) and *The Placemaker's Guide to Building Community* (2010) were written in the wake of both the post-development (Rahnema and Bawtree, 1997) and the 'tyranny of participation' (Cooke and Kothari, 2001; Hickey and Mohan, 2004a) discourses. The explicit re-contextualisation of development as perceived by Hamdi in the introduction to *The Placemaker's Guide* is valuable as a reference marker for the various transitions and shifts that methodological practices of development have attempted to transcend. He notes variously the shifting means of political articulation of development success through global macro-economic policy (Hamdi, 2010, p. 8), before confronting and contesting the challenges facing contemporary development practitioners and academics that is advocated by Browne (2007) and Schrijvers (1993).

In highlighting the practical necessity of development practice Hamdi explicitly recognises the social and material reality of the scale of global urbanisation, poverty, and malnutrition (2010, p. 8), as well as the various means by which to understand and interpret the statistical meaning of development (McGillivray and Clarke, 2006). Yet it is in this interpretation of the material, social, and practical realities of these policies at grass-roots level that we can appreciate Hamdi's attempts to analyse and communicate the methodological implications of twenty-first century development practice and identity, here quoted at length:

> We note the changing role of the expert, from lead agent to catalyst, from disciplinary to interdisciplinary work, from producing plans to cultivating opportunity. [...] We see more participation – away from sweat equity towards empowerment and power-sharing, towards partnership. [...] The development field is progressively dematerialised from shelter, water [...] to rights, governance, livelihoods. [...] There is more focus on insiders' priorities, notwithstanding the risk, which still prevails, of co-option. [...] We see a shift from practical to more strategic work in the desire to tackle root causes of poverty and to scale up programmes. [...] We move from a position of providing for the poor to enabling the poor to provide for themselves, recognising their productive capacities, reducing dependency, building resilience to the shocks and stresses of daily life. [...] We see a significant shift to urban, in view of the unprecedented growth of urban population and the strain this places on people, on resources and on the environment. Cities in the developing world will account for 95 percent of urban expansion over the next two decades and by 2030, four billion people will live in cities – 1.4 billion in slums.
>
> (Hamdi, 2010, p. 16)

Within these observations it is possible to discern clear links to our wider premise of a critical shift towards identity as a practice, with Hamdi exemplifying the practical and theoretical continuation of the premise begun by Turner's exploration of use-value and identity (1972b, pp. 152, 159). Amongst others, key phrases in the above statement bear further examination, namely the notion of 'cultivating opportunity', 'empowerment, power-sharing and partnership', 'insider's priorities', and the 'progressive demater-ialisation' of development towards 'rights, governance and livelihoods'.

Each of these various observations might be critically and thematically compared to the identity of actors, agents, and agency that engage in the act of promoting, advocating, and agitating development (Hamdi, 2010, pp. 88–89, 180). It is in his exploration of the methodological approach of participation and placemaking that Hamdi isolates the simple model of 'PEAS and the Social Side of Practice' (2010, p. 141) as a means to inculcate these ideas in both practitioners and communities alike.

PEAS, an acronym for providing, enabling, adaptability, and sustainability, describes Hamdi's ideals and activities of responsible practice in strategic action planning (SAP) (2010, p. 139). The implications of this simple re-contextualisation of development practice suggests an underlying methodological contestation of the politics of identity. Hamdi critiques traditional models of providing that are focused on *things* whilst eschewing the challenges and confrontations of people and the social context of communities (Hamdi, 2010, p. 142, 1986, p. 138; Hamdi and Goethert, 1989, pp. 20–22; Matarasso, 2007).

Such provision of *things* denies the necessity of development and place to confront and mediate the political, social, and economic relations and the principles of sustainable development.[7] Hamdi cites Marilyn Taylor's observations (2008) of similar implications before offering an alternative to what he describes as the 'paralysis of the moral and political imagination' (Illich, 1973) in the social potential of facilitating sustainable livelihoods.

For Hamdi, providing must by necessity be interlinked and interdependent with the other factors of PEAS if it is to facilitate positive social change, and resist the kind of helping that Hamdi and others recognise becomes a drug (Gronemeyer, 2010; Hamdi, 2010, p. 147). Thus *providing* only works in connection with enabling as 'the ability and willingness to provide the means to open doors and create opportunities' (Hamdi, 2010, p. 147). The interdependence of PEAS is also placed as a counter to Burgess' et al. (1997a) critique of the inevitable neoliberal co-option of development.

In focusing on the skills of development practitioners to enable and provide interactive rather than representational development Hamdi focuses on specifically small actions that become catalysts to provoke and release the positive identities of development that exist within local entrepreneurship and social relations (2010, p. 148). Once again the opportunity for critical reflection on Westernised space raised by Hamdi's observations here is striking. Perhaps the most direct example of a Western protagonist for sustainable alternative development can be found in Dan Hill's *Dark Matter and Trojan Horses*

(2015). In essence, Hill's engagement with the agency of space is twofold: firstly, engagement with the potentially positive affects of architecture and space on the social identity of the people it serves – architecture as Trojan horses (2015, p. 58); and secondly, a recognition of the need to engage with the strategic scale of political decision making and planning that underpins and defines architecture – the dark matter that surrounds architecture (2015, pp. 80–84).

Hill's most cited example is a proposed Low2No community housing development in Helsinki (2015, pp. 51–57). This project was driven by Hill's work with SITRA whose role was to act as 'an enabler of market transformation', leading to ARUP engineers and architects Saurbruch Hutton developing an innovative all timber framed building design which is currently slated for completion in 2017 (some eight years after the initial competition). For Hill however, the building itself was designed to act as an architectural McGuffin (2015, p. 50) – a vehicle whose real reason to exist is in order to move a wider more interesting narrative forwards – that would provoke a discussion of alternative architecture and development at both strategic and social levels.

Much of Hill's socio-spatial propositions reflect the same advocacies as Hamdi for engaging with interconnected strategic and community action planning in development. Whilst their shared advocacy of architecture as a catalyst for wider social change is compellingly similar, the difference between Hamdi and Hill is self-evidently the contrasting contexts in which they are working. The formal socio-political and economic landscape that defines Hill's work generates an inertia that is incredibly challenging to confront and contend.

Whilst much the same challenges face Hamdi it is reasonable to deduce that the informality and economic absence of Global South contexts offers a strikingly different (if still ridden with its own brand of complications and difficulties) context from which to engage and provoke social agency and change. Whilst the aspirations of Hill's Westernised architecture for social change are clear, he is forced to focus upon changing the political and strategic dark matter of architecture because the everyday changes needed to support such proposals are extremely unlikely to emerge from grass-roots informal spatial practices. The danger of Hill's proposals taken on their own is the potential for them to become merely another variation on formal trickle-down planning, rather than a real engagement with the positive potential of promoting socially sustainable change from within the material reality of everyday life itself.

Resonating with Hill, Hamdi reinforces his critical advocacy of small change, participatory, and grass-roots development by articulating how both practice and identity are intrinsically linked with socially sustainable development. Intersecting with discourses from Colin Ward (1996), Simon Nicholson (1972), Peter Kropotkin (2006), and Ivan Illich (1973), Hamdi's adaptability is described as integral to the process of design without necessarily producing 'an architecture of building' or 'end state'. Instead, adaptability in development practice offers a social architecture of invitation and opportunity that affords the facilitation of an open and discursive identity as a social process. In practical terms this suggests a spatial practice which represents 'a minimum of organisation that would serve the benefits of planning, while leaving individuals the greatest possible control over their lives'. It aims 'to sustain as many particularities as possible, in the hope that most people will accept, discover, or devise one that fits' (Berger and Neuhaus, 1977, p. 206).

The practical benefits of these methodological articulations of social sustainability are summarised by Hamdi's understanding of PEAS as a culture of practice that is intrinsically bound to ideas of growth, crucially coupled with mutual learning (2010, p. 151). This interpretation of knowledge and practice as forms of social sustainability involves recognition of the positive specificity of difference and subsequent identity as an aspiration of development. Yet in being explicitly a reciprocal and dialectic process between both actors and agents of development, Hamdi's socio-spatial methodologies offer a contestation of the authority, knowledge, and identity of developers themselves (2010, p. 169, 2004, pp. 125–128; Hamdi and Goethert, 1989, pp. 22–24).

The aspirational results of PEAS as a process of people building sustainable livelihoods are articulated in this interpretation of development as interdependent with facilitating spatial relations that 'reduce the dependency-inducing practices of providing as a discrete expert routine' (2010, p. 179). Hamdi explicitly recognises time and again the various 'coercive objectivity of reasoning based upon implicit principles of division, hierarchy and exclusion' (2010, p. 179) that interlink identities of dependency with development as a product, and the 'mandated empowerment' (Banerjee and Duflo, 2008) of co-option.

In contrast to this, Hamdi provides exemplars of methodological practice and outcomes ranging from a walking school bus that contests social and spatial divisions (2010, pp. 113–114), cultural centres (2010, pp. 82–83, 105–106), bus stops (2010, pp. 73–76), and even community pickling businesses (2004, p. 85), buffalo and mushroom cooperatives (2010, pp. 108–109), and entrepreneurial recycling schemes (2004, pp. 81–82). Each of these examples is dependent on the conception of development as an open and participatory process.

Noting the complicity of development with globalised 'tied aid' and the continued implications of global macro-economic policy manifestations (Hamdi, 2010, pp. 180–183; OECD, 2006), Hamdi's advocacy of the building of sustainable livelihoods as the core concern of twenty-first century development is marked in its explicit aspiration to respond to the multi-dimensional experiences of poverty (Beall and Kanjii, 1999). Livelihoods – be they informal, alternative, or largely conventional – and simple acts of sustainable social agency are the key assets and strategies that allow for first survival and then entrepreneurship that families and communities can utilise to articulate their own identities of difference through a narrative process of development (Grown and Sebstad, 1989).

Identity as critical participatory practice

In contrast to conceptions of development and identity as negative products of global ideologies, Turner's 'progressive development' (1986, p. 12) and Hamdi's 'sustainable livelihoods' (2010, p. 16) suggest far richer contestations of identity. The critical re-contextualisation of these themes reveals methodologies designed to promote the resourceful and efficient organisation of people in social and democratic participation and empowerment.

Such methodological contestations of identity disrupt the assumptions of other forms of social existence and the passive productions of development identities that are implicit within hierarchical development models (Esteva and Prakash, 1998, p. 15). They equally facilitate and invoke social, relational, and material practices that are based upon the resilience and adaptability of autonomous networks and non-hierarchical organisations of power. They produce identities and practices that are interdependently defined by the material and economic choices of people seeking to develop their own solutions to their own problems.

By engaging in such contestations of identity, alternative development practices are implicated with a radical pluralism and subjectivity that reflects a critical counter-narrative to the conventions of Western hegemony (Esteva and Prakash, 1998, p. 36). This reflects recognition and contestation of the concept of 'what is right', not as a prescription of identity, but instead as a relational and momentary condition of material practices. This in turn implies that such ideas of development are not based upon the desire for stationary and static constructions of socio-spatial relations and identity (Massey, 2005, p. 23), or 'flexible positional superiority', but must be presupposed by the inevitability of change and the necessity to reframe and question the relationships and constructed realities of such change (Bhabha, 2006, p. 156). As such Hamdi's practices and methodologies reflect an intersubjective and dialectical model of development and identity (Gillespie and Cornish, 2010), and intersect with the post-modern anthropological advocacies of Andrew Long:

> The search for inner meaning (right interpretation) only obscures or actually prevents description taking place. To seek many interventions is important in that there are always a multiplicity of meanings. By definition, a description of discourse allows for a multiplicity of truths (interpretations) that can only be revealed as they are played out in an active context.
> (1992, p. 163)

Concurrent with post-development discourse, the anthropological and sociological critique of development outlined by the 'tyranny of participation' (Cooke and Kothari, 2001) can offer a similarly valuable contextualisation of the issues of identity and the politically 'active context' of participatory development practice (Hickey and Mohan, 2004a, p. 10). This critical contestation of the assumed political neutrality of such practical participation offers similar critiques to those of post-development (Rahnema, 2010b). Yet unlike the passive theoretical critique of post-development (Hickey and Mohan, 2004b, p. 62), this discourse pursues the positive potential for participation at a grass-roots level in themes of radical democracy (Hickey and Mohan, 2004a, p. 12, 2004b, p. 59; Williams, 2004, p. 98) and transformation through social agency (Cornwall, 2004, p. 82).

Contextualising the political challenges facing participatory development in the late twentieth century, Glyn Williams notes the obscurity and complexity of global policy and external agency in the development process (2004, p. 93). He suggests that participatory practices have largely become an institutionalised process concurrent with Ferguson's 'Anti-Politics Machine' (1990). Echoing our comparative critique of development identity in the last chapter, for Williams such co-opted participatory practices came to exist as a Foucauldian exercise of power that can be interpreted as rewriting the identity of the developing world through encounters of participation, performance, and economic discipline (2004, p. 102). For Williams and Rahnema alike, participation has been institutionally articulated to legitimise power and reify beneficiaries of development as objective and abstract identities of macro-economic policies (Rahnema, 2010b, pp. 130–131).

As a counter to this predicament Williams advocates a radical re-politicisation of positive and agonistic participation and a methodological contestation of democracy as the cornerstone of development practice (2004, p. 102). Citing Whitehead and Gray-Molina (1999), Williams advocates the need to empower communities with political and spatial practices that articulate 'movements and moments' of participation which have the potential to articulate spaces which engage communities in their rights to democracy (Cornwall, 2002, p. 22). Hickey and Mohan further this in the call for participation that goes beyond the individual and local, becoming multi-scaled and strategic, thus offering a radicalised interdependence of citizenship and participation that disrupts the co-option and dependency of participation and inscribes development as a practice and identity (2004a, p. 12).

Ascribing post-structural plurality of identity as a fundamental and dynamic condition within a post-development participatory framework suggests an implicit confrontation of inequality as an inevitable context of development (Bhabha, 2004, pp. 34–35; Esteva and Prakash, 1998, pp. 159–160; Hickey and Mohan, 2004b, pp. 166–168). By confronting this, the discourse and agency of truly plural, political, and participatory development practice acts to empower the contestation of identity as a continually evolving idea. Subsequently Turner's and Hamdi's participatory practice can be recognised as a process of facilitating social, economic, and spatial relations that disrupt and contest the geometries of power (Massey, 2005, p. 83) and seeks, as Norman Long notes here, to transform them:

> Intervention is an ongoing transformational process that is constantly reshaped by its own internal organisations and political dynamic and by

the specific conditions it encounters and or itself creates, including the responses and strategies to local and regional groups who may struggle to define and defend their own social spaces, cultural boundaries and positions within the wider power field.

(1992, p. 37)

Both the strategic heteronomous practices of Turner, and the reflexive learning participatory practices of Hamdi reflect comparable political and spatial articulations of the critiques levelled at participatory co-option. Both works are intimately concerned with development as a process as noted by the progressive and open-ended nature of their empowerment toward economic and social sustainability (Hamdi, 2010, p. 154; Turner, 1976, p. 11).

For Turner, participation was always a question of 'whose participation in whose decisions?' (1976, p. 127) and his practice sought to strategically provide as many choices and opportunities for individuals to identity their own path towards development as possible (1972b, p. 174). The autonomy and heteronomy provided by open and non-hierarchical models of planning offered a complexity and mixture of development patterns that exploited Geddes' systems theory (1949) and Ashby's law requisite variety (1956, pp. 202–208). The richness and multiplicity of identity born from informal spatial relations is both found and practised by facilitating a similarly rich multiplicity of freedoms, choices, and options that reflect the economical, material, and social complexities of real space (Turner, 1972b, p. 164).

The more local and grass-roots approach advocated by Hamdi is anchored by a process of targeted and agonistic agitation of local contexts and conditions (2010, p. 70). His articulation of identity is far closer to an anthropological or ethnographic human and spatial approach than Turner's systematic approach. This specific focus upon the contestation of identities of vulnerability as a catalyst for ethical grass-roots development is exemplified by the human scale interactions that cast Hamdi as an interface and partner in acts of small sustainable social (and scaleable) change (2010, p. 16). Hamdi's practice reinforces the practical and theoretical implications of notional identities of vulnerability and superiority as key elements of a spatial practice based upon situated analysis and relational intervention (Cornwall, 2004, pp. 78–79). In comparison with the post-modern anthropology, Gardner and Lewis raise this issue explicitly with the notion of targeting a:

'relational' view of social and economic life, which stresses the interdependent but conflictual sets of relations which make up communities. [...] What holds the targeting idea together is the objective of including people who have been 'left out' of the development process.

(Gardner and Lewis, 1996, p. 106)

In targeting identities of vulnerability that exist within informal communities and sites of peripheral development, Hamdi seeks to expose the spatial,

social, and political dynamics of poverty. His focus upon tackling the implications of intergenerational transmission of identities of inequity as 'often
rooted in cast, clan and engendered cultural norms' (2010, p. 53) advocates a
confrontation with the questions of 'mutuality and identity' at a practical
grass-roots level (2010, p. 154). Thus Hamdi expressly acknowledges the
complexity and specificity of vulnerability, noting that:

> Vulnerability, however, particularly when targeting its root causes is pro-
> blematic in various ways. [...] First how do we draw boundaries around a
> condition that is constantly changing where people go in and out of being
> vulnerable – and in a globalised world, where risk may be induced in one
> place and vulnerability experienced in another?
>
> (Hamdi, 2010, p. 154)

This critical and unceasing awareness of the relationality of vulnerability,
target groups, identity, and the 'right questions' are merely the surface of the
true complexity that faces post-development and 'transformative participatory'
practice. Yet ultimately, discourses of post-development and critical participatory practice reframe development identity at the central disjunction of
action and practice, authority, and identity. Thus the expectation that development can ever be right, or can ever work for the right people or target the
right issues as a means to solve the assumed problems of inequality is in itself
a crucial characteristic of the subjective narrative facing development:

> Patterns of social differentiation then are only made meaningful when
> situated in terms of everyday social practices and situations. In other
> words, it is necessary to show how relationships, resources and values are
> contextualised (actualised) through specific action contexts, and the focus
> on action is central to the endeavor.
>
> (N. Long, 1992, p. 164)

Such inequity of identity as a construct of geometries of power is globally
prevalent irrespective of economic or political context. Yet in facing the
structural and subjective implausibility of identities of mutual equality, both
Turner and Hamdi are able to propose methodologies that advocate the specificity and plurality of space, and contestations of mutuality in the practice of
development and identity.

Identity and development as interdependent practices

In this chapter we have sought to explore a critical comparison to identity as
a historical, political, and theoretical and physical product in identities of
'otherness' (Robbins et al., 2008, p. 1077) and 'under-development' (Esteva
et al., 2013, p. 5). In comparing Turner's and Hamdi's discourses against
interdisciplinary intersections with post-structural politics, economics, and

identity this analysis provides a new way to interpret their respective works and the broader surrounding discourses of development, post-colonialism, and participation. Their advocacy of practical, cultural, and materially contextual approaches to development reveal the positive potential and necessity of multiplicity and (sub)alterity of identity.

The interdisciplinary contextualisation against Edward Said's critique of orientalism provides a comparable intersection of identity and 'flexible position authority' in historical empire and colonialism. The post-colonial context in which both Turner's and Hamdi's discourses are engaged is implicitly a contestation of the continued implications of negative differentiations of identity outlined by Said. This critical analysis of identity as a means of producing change and controlling space allows an insight into global patterns of historical intervention. Such an analysis provides an interpretation of political methodologies as successively seeking the control and manipulation of the economically and culturally peripheral *other*, by exploiting the authority of the centre for coercive political and economical purposes.

The subsequent implications of post-colonial and post-development discourse suggest a re-contextualisation of identity that articulates a theoretical deconstruction of semantically and politically prescriptive forms of formal authority towards a pluralism and multiplicity of development. Thus, our contention remains that in facilitating the positive potential of alternative development methodologies, Turner and Hamdi each offer practical examples of the implications of post-structural and post-colonial plural identities and social trajectories.

In re-articulating an open and non-prescribed notion of identity, the values and needs of specific local community agendas become the bedrock of the empowerment of participatory democracy. This re-contextualisation allows us to look at such development as truly plural and built upon an open multiplicity of identity as interdependent with socio-cultural specificity and relations that encompass more than abstract economic criteria. The interwoven social and spatial relations of production, exchange, and consumption are understood and valued for their interdependence with the complex network of local social, political, and cultural contexts (Turner, 1976, p. 17). This alternative articulation of what development *means* and *does* intersects with the inherent complexity, richness, and agonism of activities with competing representations and interpretations concerning ideology, identity, power, and knowledge that transcend any static representation of space as merely an economic product (Laclau and Mouffe, 2001, p. 183).

The notion of both development and concurrently identity as practice suggests an explicit social democratic and grass-roots engagement with political empowerment, self-narration, and identity. Thus, the potential for such empowerment is richly evidenced in Hamdi's encounters with enterprising and insightful community protagonists who act as social agents who disrupt, inspire, and provoke their own community to explore and define their own identity and development trajectory:

Development, he said, happens when people, however poor in money, get together, get organised, become sophisticated and go to scale. It happens when they are savvy and able to influence and change the course of events or the order of things locally, nationally or even globally – or are themselves able to become that order or part of it. Development, he said, is that stage you reach when you are secure enough in yourself, individually or collectively, to become interdependent; when 'I' can emerge as 'we', and also when 'we' is inclusive of 'them'.

(Hamdi, 2004, p. v)

Outside the critical comparisons put forward in this book neither Turner's or Hamdi's contributions to development and cultural identity theory have been previously examined in the context of wider socio-spatial or development theory. Yet by re-reading and re-framing the disarmingly simplistic texts of both protagonists, this chapter has highlighted unconscious appropriations of post-structural concepts of identity and development theory that offer new hope and potential for development practitioners (McKinnon, 2006, p. 22). And consequently, just as in the reflective identity of orientalism, they provide a provocative challenge to Western ideological notions of ontological authority and stability that transcend the humble origins of post-development practice:

In so doing, we enable people to find new ways of doing, thinking and relating in response to everyday problems which one takes for granted – breaking down barriers; optimising not maximising. These are the qualities of leadership in practice and for development – a new openness to dialogue and learning.

(Hamdi, 2004, p. xiv)

These comparisons provide a framework from which to re-assess and re-contextualise both the theoretical discourses of orientalism and post-colonialism, and the practical discourses of Turner and Hamdi. The inherently negative and critical discourse of orientalism and post-colonial studies are thus provided with alternative positive contextualisations when compared with the practical methodologies of Turner and Hamdi.

Reflexively, Turner's and Hamdi's respective discourses can be con-textualised and read anew as contestations of contemporary discourses of identity that exist within a broad interdisciplinary academia. This provocative comparison allows a renewed advocacy of the need to materially and practically contest theoretical discourse in order to avoid the dramatic implications of spatial, academic, and political abstraction. This interdisciplinary contextuali-sation of Turner and Hamdi is a small step towards an open and positive engagement with the interdependence of cultural identity with both development and architectural spatial practices.

In seeking to critically compare the full positive potential of a comparison with the development methodologies of Turner and Hamdi, the analysis here highlights exemplars of socially and economically sustainable spatial relations and practices; examples of the interdependence of identity and practice as crucial in any notional re-imagining of architecture and space in both the Global South and Global North. And these thematic contestations of the interdependent relationship of identity and practice are the subject of the next chapter in this book, as we look to see how identity and practice can be engaged in the pursuit of an alternative engagement with the interdependence of identity, space, and architecture.

Notes

1 Such as the relationships and identities that were used to negatively define women, the insane, criminals, and the poor. This analysis would become crucial to the work of key post-structural theorists such as Foucault (2006, 1998, 1991a, 1991b).
2 As highlighted in his new preface to the 2003 edition of *Orientalism*.
3 It should be noted that whilst the point is conceded to Warraq, the vehement tone of the attack is also considered largely unnecessary and unhelpful. It is also noted that the critiques of Irwin are equally intriguing in their expansion of orientalism into German and Russian contexts, however they similarly are tangential to the utilisation of Said's core discourse in our research (2007).
4 Various examples can be explored here including the continued contemporary military and economic manifestations of Western government and corporations in the Middle East, the political interventions in the ongoing Syrian conflict, as well more historical examples such as the Indian independence movement and various Western interventions in South American political processes.
5 Notably from Turner's experiences, abstract assumptions prevail even when practice is approached with the best of intentions.
6 A premise which could similarly be construed in the context of the homogenous model of development that underpins Westernised space, and devalues, dismisses, or destroys examples of alternative development, for example the plotlander movement in the UK.
7 This analysis parallels contemporary Western discussion concerning the political role of the architecture profession, highlighted in 2014 by the ongoing purposefully antagonistic and media seeking contributions of Zaha Hadid and Patrick Schumacher (Lambert, 2014; Reich, 2014).

References

Abdilahi Bulhan, H., 1985. *Frantz Fanon and the Psychology of Oppression*. Plenum Press, New York.

Ashby, W., 1956. *An Introduction to Cybernetics*. Chapman and Hall, London.

Ashcroft, B., Griffiths, G., Tiffin, H., 2003. *The Post-Colonial Studies Reader*, 2nd Edition. Routledge, London.

Attridge, D., Bennington, G., Young, R.J.C. (Eds), 1989. *Post-Structuralism and the Question of History*. Cambridge University Press, Cambridge.

Banerjee, A., Duflo, E., 2008. *Mandated Empowerment: Handing Anti-Poverty Policy Back to the Poor?* New York Academy of Science, New York.

Bauman, Z., 2000. *Globalization: The Human Consequences*. Columbia University Press, New York.

Beall, J., Kanjii, N., 1999. *Households, Livelihoods and Urban Poverty*. Theme Paper 3. International Development Department, University of Birmingham, Birmingham, UK.

Berger, P.L., Neuhaus, R., 1977. *To Empower People: The Role of Mediating Structures in Public Policy*. American Enterprise Institute for Public Policy Research, Washington, DC.

Bhabha, H.K., 1990. *Nation and Narration*. Routledge, London.

Bhabha, H.K., 2004. *The Location of Culture*. Routledge, London.

Bhabha, H.K., 2006. Cultural Diversity and Cultural Differences, in: Ashcroft, B., Griffiths, G., Tiffin, H. (Eds), *The Post-Colonial Studies Reader*. Routledge, New York.

Browne, S., 2007. *Aid and Influence: Do Donors Help or Hinder?*Earthscan, London.

Burgess, R., 1977. Petty Commodity Housing or Dweller Control? A Critique of John Turner's Views on Housing Policy. *World Development* 6, 1105–1133.

Burgess, R., 1978. Self-Help Housing. A New Imperialist Strategy? A Critique of the Turner School. *Antipode* 9, 50–60.

Burgess, R., Carmona, M., Kolstee, T. (Eds), 1997a. *The Challenge of Sustainable Cities: Neoliberalism and Urban Strategies in Developing Countries*. Zed Books, London.

Burgess, R., Carmona, M., Kolstee, T., 1997b. Contemporary Spatial Strategies and Urban Policies in Developing Countries: A Critical Review, in: *The Challenge of Sustainable Cities*. Zed Books, London.

Burgess, R., Carmona, M., Kolstee, T., 1997c. Contemporary Macroeconomic Strategies and Urban Policies in Developing Countries: A Critical Review, in: *The Challenge of Sustainable Cities*. Zed Books, London.

Burnell, J., 2012. Small Change: Understanding Cultural Action as a Resource for Unlocking Assets and Building Resilience in Communities. *Community Development Journal* 48, 134–150.

Chang, H.-J., 2002. *Kicking Away the Ladder: Development Strategy in Historical Perspective: Policies and Institutions for Economic Development in Historical Perspective*. Anthem Press, London.

Clifford, J., 1980. *Orientalism* – Book Review. *History and Theory* 19(2), 204–223.

Cooke, B., Kothari, U. (Eds), 2001. *Participation: The New Tyranny?*Zed Books, London.

Cornwall, A., 2002. *Making Spaces, Changing Places: Situating Participation in Development*, Working Paper series 170. IDS, Brighton.

Cornwall, A., 2004. Spaces for Transformation? Reflections on Issues of Power and Difference in Participation in Development, in: Hickey, S., Mohan, G. (Eds), *Participation: From Tyranny to Transformation*. Zed Books, London.

Eagleton, T., 2006. Lust for Knowing – Book Review. *New Statesman*, 13 February.

Eagleton, T., Jameson, F., Said, E. (Eds), 2001. *Nationalism, Colonialism, and Literature*, 5th Edition. University of Minnesota Press, Minneapolis.

Edelman, M., 1999. *Peasants Against Globalization: Rural Social Movements in Costa Rica*. Stanford University Press, Stanford, CA.

Escobar, A., 2010. Planning, in: Sachs, W., *The Development Dictionary*. Zed Books, London.

Esteva, G., 2010. Development, in: Sachs, W., *The Development Dictionary*. Zed Books, London.

Esteva, G., Prakash, M.S., 1998. *Grassroots Post-Modernism: Remaking the Soil of Cultures*. Zed Books, London.

Esteva, G., Babones, S.J., Babcicky, P., 2013. *The Future of Development: A Radical Manifesto*. Polity Press, Cambridge.

Eversole, R., 2012. Remaking Participation: Challenges for Community Development Practice. *Community Development Journal* 47, 29–41. doi:10.1093/cdj/bsq033

Fabian, J., 2002. *Time and the Other*. Columbia University Press, New York.

Fanon, F., 2001. *The Wretched of the Earth*, Reprinted Edition. Penguin Classics, London.

Ferguson, J., 1990. *The Anti-Politics Machine: 'Development', Depoliticisation, and Bureaucratic Power in Lesotho*. Cambridge University Press, Cambridge.

Ferguson, J., Gupta, A., 1992. Beyond 'Culture': Space, Identity and the Politics of Difference. *Cultural Anthropology* 7, 6–23.

Fichter, R., Turner, J.F. (Eds), 1972. *Freedom to Build*. Macmillan, New York.

Foucault, M., 1991a. Governmentality, in: *The Foucault Effect: Studies in Governmentality*. University of Chicago Press, Chicago.

Foucault, M., 1991b. *Discipline and Punish: The Birth of the Prison*, New Edition. Penguin, London.

Foucault, M., 1998. *The History of Sexuality: The Will to Knowledge*, New Edition. Penguin, Harmondsworth.

Foucault, M., 2006. *Madness and Civilization*. Vintage Books, New York.

Gardner, K., Lewis, D., 1996. *Anthropology, Development and the Post-Modern Challenge*. Pluto Press, London.

Geddes, P., 1949. *Cities in Evolution*, 2nd Edition. Williams and Norgate, London.

Gillespie, A., Cornish, F., 2010. Intersubjectivity: Towards a Dialogical Analysis. *Journal for the Theory of Social Behaviour* 40, 19–46.

Gordon, L., 1995. *Fanon and the Crisis of European Man*. Routledge, New York.

Grenell, P., 1972. Planning for Invisible People: Some Consequences of Bureaucratic Values and Practices, in: Fichter, R., Turner, J.F. (Eds), *Freedom To Build*. Macmillan, New York.

Gronemeyer, M., 2010. Helping, in: Sachs, W., *The Development Dictionary*. Zed Books, London.

Grown, C., Sebstad, J., 1989. Introduction: Towards a Wider Perspective on Women's Employment. *World Development* 17(7), 937–952.

Guha, R., Spivak, G.C., 1989. *Selected Subaltern Studies*. Oxford University Press, New York.

Hall, S., Massey, D., Rustin, M., 2013. After Neoliberalism: Analysing the Present. *Soundings* 53 (April), 8–22.

Hamdi, N., 1986. Training and Education: Inventing a Programme and Getting it to Work. *Habitat International* 10, 131–140.

Hamdi, N., 2004. *Small Change*. Earthscan, London.

Hamdi, N., 2010. *The Placemaker's Guide to Building Community*. Earthscan, London.

Hamdi, N., Goethert, R., 1989. The Support Paradigm for Housing and its Impact on Practice: The Case in Sri Lanka. *Habitat International* 13, 19–28.

Harris, R., 2003. A Double Irony: The Originality and Influence of John F.C. Turner. *Habitat International* 27, 245–269.

Harvey, D., 1996. *Justice, Nature and the Geography of Difference*. Blackwell Publishing, Oxford.

Harvey, D., 2015. *Seventeen Contradictions and the End of Capitalism*. Profile Books, London.

Hickey, S., Mohan, G. (Eds), 2004a. *Participation: From Tyranny to Transformation.* Zed Books, London.

Hickey, S., Mohan, G., 2004b. Relocating Participation Within a Radical Politics of Development: Insights from Political Practice, in: *Participation: From Tyranny to Transformation.* Zed Books, London.

Hill, D., 2015. *Dark Matter and Trojan Horses: A Strategic Design Vocabulary.* Strelka Press, Moscow.

Illich, I., 1973. *Tools for Conviviality.* Calder and Boyars, London.

Illich, I., 2010. Needs, in: Sachs, W., *The Development Dictionary.* Zed Books, London.

Irwin, R., 2007. *For Lust of Knowing: The Orientalists and Their Enemies.* Penguin, London.

Jameson, F., 2001. Modernism and Imperialism, in: Eagleton, T., Jameson, F., Said, E. (Eds), *Nationalism, Colonialism and Literature.* University of Minnesota Press, Minneapolis.

Kiely, R., 1994. Development Theory and Industrialisation: Beyond the Impasse. *Journal of Contemporary Asia* 24, 133–160.

Kiely, R., 1999. The Last Refuge of the Noble Savage? A Critical Assessment of Post-Development Theory. *The European Journal of Development Research* 11, 30–55.

Kropotkin, P., 2006. *Mutual Aid: A Factor of Evolution, Dover Books on History, Political and Social Science.* Dover, New York.

Laclau, E., Mouffe, C., 2001. *Hegemony and Socialist Strategy.* Verso, London.

Lambert, L., 2014. Open-Letter to Mr Patrik Schumacher: Yes, Architects are Legitimised and Competent to Address the Political Debate. *The Funambulist.* Available at http://thefunambulist.net/2012/02/02/architectural-theories-open-letter-to-mr-patrick-schumacher-yes-architects-are-legitimized-and-competents-to-address-the-political-debate/ (accessed 15 March 2016).

Latouche, S., 2010. Standards of Living, in: Sachs, W., *The Development Dictionary.* Zed Books, London.

Lefebvre, H., 2009. *Henri Lefebvre – State, Space, World: Selected Essays.* University of Minnesota Press, Minneapolis.

Lewis, H.S., 2007. The Influence of Edward Said and Orientalism on Anthropology, or: Can the Anthropologist Speak? *Israel Affairs* 13(4), 774–785.

Long, A., 1992. Goods, Knowledge and Beer; The Methodological Significance of Situational Analysis and Discourse, in: Long, N., Long, A. (Eds), *Battlefields of Knowledge.* Routledge, London.

Long, N., 1992. From Paradise Lost to Paradigm Regained?; The Case for an Actor-Oriented Sociology of Development, in: Long, N., Long, A. (Eds), *Battlefields of Knowledge.* Routledge, London.

Lummis, C.D., 1997. *Radical Democracy.* Cornell University Press, Ithaca, NY.

Lummis, C.D., 2010. Equality, in: Sachs, W., *The Development Dictionary.* Zed Books, London.

Mann, W., 2003. The Plotlands Experience: The Self-Build Settlements of Southeast England. *Oase* 61 (Spring), 110–123.

Massey, D., 1994. *Space, Place and Gender.* Polity Press, Cambridge.

Massey, D., 2004. Geographies of Responsibility. *Geografiska Annaler Series B: Human Geography* 86, 5–18.

Massey, D., 2005. *For Space.* Sage Publications, London.

Matarasso, F., 2007. Common Ground: Cultural Action as a Route to Community Development. *Community Development Journal* 42, 449–458.

Mathey, K., 1991. *Beyond Self-Help Housing*. Mansell, London.

McGillivray, M., Clarke, M. (Eds), 2006. *Understanding Human Well-being*. United Nations University Press, Tokyo.

McKinnon, K.I., 2006. An Orthodoxy of 'The Local': Post-Colonialism, Participation and Professionalism in Northern Thailand. *The Geographical Journal* 172, 22–34.

Moore-Gilbert, B., 2000. Spivak and Bhabha, in: Ray, S., Schwartz, H. (Eds), *A Companion to Postcolonial Studies*. Blackwell Publishing, Oxford.

Nandy, A., 2002. The Beautiful Expanding Future of Poverty: Popular Economics as a Psychological Defense. *International Studies Review* 4(2), 107–221.

Nicholson, S., 1972. The Theory of Loose Parts. *Studies in Design Education and Craft Technology* 4, 5–14.

OECD, 2006. *Development Cooperation Report 7*. OECD, Washington, DC.

Parry, B., 2006. Problems in Current Theories of Colonial Discourse, in: Ashcroft, B., Griffiths, G., Tiffin, H. (Eds), *The Post-Colonial Studies Reader*. Routledge, New York.

Peattie, L., 1982. Some Second Thoughts on Sites and Services. *Habitat International* 6 (1–2), 131–139.

Peattie, L., Doebele, W., 1973. Review of Freedom to Build. *Journal of the American Institute of Planning* 39, 66–67.

Rahnema, M., 2010a. Poverty, in: Sachs, W., *The Development Dictionary*. Zed Books, London.

Rahnema, M., 2010b. Participation, in: Sachs, W., *The Development Dictionary*. Zed Books, London.

Rahnema, M., Bawtree, V. (Eds), 1997. *The Post-Development Reader*. Zed Books, London.

Reich, J., 2014. Zaha Hadid Defends Qatar World Cup Role Following Migrant Worker Deaths. *The Guardian*, 25 February. Available at http://www.theguardian.com/world/2014/feb/25/zaha-hadid-qatar-world-cup-migrant-worker-deaths (accessed 15 March 2016).

Rist, G., 2006. *The History of Development*, 3rd Edition. Zed Books, London.

Robbins, B., 1993. The East as a Career, in: Sprinkler, M. (Ed.), *Edward Said: A Critical Reader*. Wiley-Blackwell, Chichester.

Robbins, S., Cornwall, A., Von Lieres, B., 2008. Rethinking 'Citizenship' in the Postcolony. *Third World Quarterly* 29, 1069–1086.

Rodell, M.J., Skinner, R.J., 1983. Introduction. Contemporary Self-help Programmes, in: *People, Poverty and Shelter. Problems of Self-Help Housing in the Third World*. Methuen, London.

Roy, A., 2011. Slumdog Cities: Rethinking Subaltern Utopianism. *International Journal of Urban and Regional Research* 35, 223–238.

Rubin, A.N., 2003. Techniques of Trouble: Edward Said and the Dialectics of Cultural Philology. *South Atlantic Quarterly* 102, 862–876.

Ruthven, M., 2003. Obituary: Edward Said. *The Guardian*, 26 September. Available at http://www.theguardian.com/news/2003/sep/26/guardianobituaries.highereducation (accessed 15 March 2016).

Sachs, W., 2010a. Preface to the New Edition, in: *The Development Dictionary*. Zed Books, London.

Sachs, W., 2010b. *The Development Dictionary. A Guide to Knowledge as Power*, 2nd Edition. Zed Books, London.

Sachs, W., 2010c. One World, in: *The Development Dictionary*. Zed Books, London.

Said, E., 1983. *The World, the Critic, and the Text*. Harvard University Press, Cambridge, MA.

Said, E., 2001. Yeats and Decolonisation, in: Eagleton, T., Jameson, F., Said, E. (Eds), *Nationalism, Colonialism and Literature*. University of Minnesota Press, Minneapolis.

Said, E., 2003. *Orientalism*. Penguin Classics, London.

Schrijvers, J., 1993. *The Violence of Development*. International Books, Utrecht.

Sennett, R., 2004. *Respect: The Formation of Character in an Age of Inequality*. Penguin, London.

Slemon, S., 2006. The Scramble for Post-colonialism, in: Ashcroft, B., Griffiths, G., Tiffin, H. (Eds), *The Post-Colonial Studies Reader*. Routledge, New York.

Spivak, G.C., 1998. Can the Subaltern Speak?, in: Grossberg, L., Nelson, C. (Eds), *Marxism and the Interpretation of Culture*. Macmillan Education, Basingstoke.

Spivak, G.C., 1999. *A Critique of Postcolonial Reason: Toward a History of the Vanishing Present*. Harvard University Press, Cambridge, MA.

Spivak, G.C., 2004. Righting Wrongs. *South Atlantic Quartlery* 103, 523–581.

Taylor, M., 2008. *Transforming Disadvantaged Places: Effective Strategies for Place and People*. Joseph Rowntree Foundation, York.

Truman, H.S., 1967. *Inaugural Address*. Documents on American Foreign Relations, Princeton University Press, Princeton, NJ.

Turner, J.F.C., 1963. Dwelling Resources in South America. *Architectural Design* 8, 360–393.

Turner, J.F.C., 1967. Barriers and Channels for Housing Development in Modernizing Countries. *Journal of the American Institute of Planning* 33, 167–181.

Turner, J.F.C., 1968a. Uncontrolled Urban Settlements: Problems and Policies. *International Social Development Review* 1, 107–128.

Turner, J.F.C., 1968b. Housing Priorities, Settlement Patterns, and Urban Development in Modernizing Countries. *Journal of the American Institute of Planners* 34, 354–363.

Turner, J.F.C., 1972a. The Re-Education of a Professional, in: Fichter, R., Turner, J.F. (Eds), *Freedom To Build*. Macmillan, New York.

Turner, J.F.C., 1972b. Housing as a Verb, in: Fichter, R., Turner, J.F. (Eds), *Freedom To Build*. Macmillan, New York.

Turner, J.F., 1974. The Fits and Misfits of People's Housing. *RIBA Journal* 81(2), 12–21.

Turner, J.F., 1976. *Housing by People: Towards Autonomy in Building Environments*. Marion Boyars, London.

Turner, J.F., 1983. From Central Provider to Local Enablement. *Habitat International* 7, 207–210.

Turner, J.F., 1986. Future Directions in Housing Policy. *Habitat International* 10, 7–25.

Ward, C., 1996. *Talking to Architects*. Freedom Press, London.

Ward, C., Hardy, D., 1972. Plotlanders. *Oral History Journal* 13, 57–70.

Warraq, I., 2007. *Defending the West: A Critique of Edward Said's 'Orientalism'*. Prometheus, New York.

Whitehead, L., Gray-Molina, G., 1999. *The Long-Term Politics of Pro-Poor Policies.* World Bank. Available at http://siteresources.worldbank.org/INTPOVERTY/Resources/WDR/DfiD-Project-Papers/whitehea.pdf (accessed 14 March 2016).

Williams, G., 2004. Evaluating Participatory Development: Tyranny, Power and (Re)Politicisation. *Third World Quarterly* 25, 557–578.

Young, I.M., 2011. *Justice and the Politics of Difference*, New Edition. Princeton University Press, Princeton, NJ.

Young, R., 1990. *Untying the Text: A Post-structuralist Reader*, New Edition. Routledge, London.

Young, R., 2004. *White Mythologies: Writing History and the West*, 2nd Edition. Routledge, London.

5 Coevalness, textuality, and critical spatial practice

Previously we have explored the interdependence of disruption and social change in our comparisons of Hamdi and Massey, and the development history and critical emergence of identity as a practice. Building upon the analysis of these themes this chapter will seek to synthesise and advance these ideas further by critically re-contextualising Hamdi's participatory development methodologies against the ambiguity of post-structural identities and textual value.

As we explored in the previous chapter, contemporary development practice in the post-colonial Global South offers novel opportunities for critical comparison with key aspects of socio-spatial theory, and Westernised space more broadly. Such informal spaces exist outside Westernised conventions yet are also subject to the effects of global capitalist ideologies. And it is these contradictions that provide the opportunity to learn from exemplars of participatory development and compare them to the conditions of Western space.

In our attempts to explore a re-imagining of Westernised architecture and space we will explore a critical re-reading and comparison of Hamdi's participatory methodologies against notions of multiplicity, coevalness, and interpretations of spatial hybridity drawn from post-structural theory. In doing so we will examine Hamdi's critique of the 'forward reasoning' of institutionalised contemporary development, and his alternative advocacy of 'backward reasoning'. This distinction will help to reveal and examine the complexity and ambiguity of the social and spatial conception of value in Hamdi's development practices. Subsequently, this analysis of Hamdi's engagement with the practical and material reality of participatory development allows us to propose speculative connections to the post-structural and post-colonial observations of Doreen Massey, Johannes Fabian, Homi K. Bhabha and Gayatri Spivak.

Hamdi's methodologies will first be framed against the advocacy of postmodern anthropology theorists and practitioners that field-work and ethnography be understood as practices of coevalness (Massey, 2005, pp. 79–80). Here a return to Massey is crucial. Her analysis will offer a spatial re-contextualisation of Fabian's advocacy of 'the coevalness of time' (Fabian, 2002, p. 69). Appropriated from anthropological theory coevalness can be re-read and re-framed against overtly spatial methods of 'situated analysis' (Gardner and Lewis, 1996, p. 158) and 'embedded material practices' (Long and Long, 1992, p. 164). This chapter proposes that key aspects of Hamdi's participatory methodologies such as 'backward reasoning' and 'open learning' can be recognised as reflecting Massey's advocacy of the political specificity and relationality of space (2005, pp. 100–103) by engaging in practical methodologies and realisations of Fabian's anthropological advocacy of 'coevalness'.

Subsequently, Hamdi's articulation and methodological insight into the spatial implications of practice as a monologue or dialogue confronts the implications of post-colonial, coeval, and textual values in the pursuit of positive multiplicity of space. This analysis of Hamdi's methodological advocacy of the necessary physical proximity, human interaction, and engagement with material and social reality of development is vital. We will see that it informs spatial methodologies that invoke and empower alternative perceptions of communication that, once again, reflect key concepts of dialogue and negotiation drawn from post-structural theory.

Hamdi's methodological tropes of coeval, reflexive, and open communication are seen as a practical means to contest, challenge, and empower community participation, and frames our comparisons with Homi K. Bhabha's theoretical discourse on the creation of a hybrid 'third-space'. Bhabha's concepts of 'third-space' and 'enunciations of meaning' (2004, p. 254) reveals that the implicit negotiation of alternative and hybrid identities and values that are explored implicitly in Hamdi's methodologies. This intersection of theoretical and practical engagements with the enunciation of alternative identity and values allows for the final trajectory revealed in our research; the comparison of Hamdi's work with Spivak's discourses on subaltern theory (1990, pp. 225–228).

Hamdi's advocacy of spatial methodologies of spontaneity (2004, p. 98), open-ended practice, and an interconnected mutual learning reveal an approach to development as an act of both learning from and empowering people through dialogue (2010, p. 175). This critical approach necessitates finding ways to engage with (informal) communities in open, self-reflective, and creative ways that create opportunities for people to define their own values and control their own spatial and social relations of development. In essence to not only produce their own space, but their own identities and social relationships in a way that does not (necessarily) conform to outside expectations or cultural assumptions of universal Westernised development identities and values. This analysis proposes the grass-roots development practice methodologies of Nabeel Hamdi as realisations of the textualised and enunciated values, advocated by Spivak as 'an ethical kind of reading attentive to the aporetic structure of "knowing" in the encounter with the other' (Morris, 2010, p. 9).

When critically compared with Bhabha's notions of ambiguity, enunciation, and textual values, or Spivak's Derridean deconstruction of Marxist 'materialist subject' values (1984, p. 232), Hamdi's methodologies must be re-read and re-valued. His approach to development can be re-interpreted as an example of pluralist, post-colonial space, and values that (as we have explored in previous chapters) have been long advocated by Western spatial theorists, yet seldom realised positively and sustainably in practice. Paradoxically, the achievement of such spaces perhaps suggests an inversion of many meta-narratives of Western spatial practices and their underlying economic models. Spivak offers a glimpse of these dialectic implications in a disruption of Marx's notions of post-colonial labour theory and post-structural values against global 'shifting lines' and the 'dark presence of the third world' (1985, p. 84).

This chapter's contention that the enunciated re-articulation of Hamdi's participatory development can be interpreted as an exemplar of informal dialectic and materialist practice thus suggests a re-articulation of our narrow understanding of the relationship between the social production, enunciation, and everyday experience of value. It thus also provides a final and invaluable strand of critical thought that is needed as we seek to re-imagine Western space and architecture by learning from difference, multiplicity, and otherness.

Multiplicity

The theoretical complexity of the comparisons and analysis proposed by this chapter necessitate the return here to key spatial concepts articulated by Massey. As we have touched upon at various points in this book, *For Space* (Massey, 2005) offers a broad and layered critical analysis of the positive implications of conceiving space and time as interdependent. In response to the complexity of Massey's conception of the 'relational production of space', in this chapter we return once again to the implications of the multiplicity of space in comparison to the concrete practical realisations of Hamdi's development methodologies.

In previous chapters we have explored how the notion of multiplicity in space explicitly connects to the perception and practice of development. The implications of space as a multiplicity confronts the inherent political, economic, and spatial relations that define the historical and contemporary conditions of inequality that underpin development practice: necessity, scarcity, and absence. These themes of economic, political, and social absence have begun to be engaged with in Westernised contexts by acacemics like Jeremy Till at SCIBE (2012), but the idea of scarcity remains a highly under-examined aspect of research not focused explicitly on the Global South.

In contrast, Massey's theoretical exposition of various theoretical mis-conceptions of space demonstrated in both structuralism (2005, pp. 36–42) and post-structuralism (2005, p. 158, 2004, p. 9), provides a series of profound re-contextualisations of the positive potential of space to provide a foundation for a re-imagining of alternative development and values:

> First, that we recognise space as the product of interrelations: as con-stituted through interactions, from the immensity of the global to the intimately tiny. Second, that we understand space as the sphere of the

possibility of the existence of multiplicity in the sense of contemporaneous plurality: as the sphere in which distinct trajectories coexist: as the sphere therefore of coexisting heterogeneity. Without space, no multiplicity: without multiplicity, no space. If space is indeed this product of inter-relations, it must be predicated upon the existence of plurality. Multiplicity and space as co-constitutive. Third, that we recognise space as always under construction. Precisely because space on this reading is a product of relations-between, relations which are necessarily embedded material practices which have to be carried out, it is always in the process of being made. It is never finished; never closed. Perhaps we could imagine space as a simultaneity of stories-so-far.

(Massey, 2005, p. 10)

Whilst the first two themes Massey notes here have been explored in our previous chapters, Massey's third observation and advocacy of understanding spatial relations as 'embedded material practice' and the recognition that such practices are 'never finished' and 'never closed' marks the origin of the comparisons in this chapter of multiplicity and development practice. These interdependent notions of spatial proximity and open-ended practice can be clearly traced throughout the comparisons of development practice outlined in this book, notably the thematic foundations of Lefebvre's spatialised material dialectics, Massey's analogy of 'space as a simultaneity of stories-so-far' (2013, p. 266), and Hamdi's participatory practices of disruption and social catalysis.

Massey's spatial appropriation and articulation of multiplicity acknowledges and engages with the relationality and specificity of spatial and political practices (2005, p. 100). Thus, if space is an open and rich multiplicity it must be recognised as constructed of both interdependent specificity and relationality. Or, in other words, place is locally unique but that local specificity is produced through its socio-spatial relations to other places both local and global. Crucially therefore the positive and rich multiplicity of places is only realised through practices in space which connect, react, and change the political, economic, and cultural relations that are infused in local and global spatial relations:

If space is rather a simultaneity of stories-so-far, then places are collections of those stories, articulations within the wider power-geometries of space. Their character will be a product of these intersections within that wider setting, and of what is made of them. And, too, of the non-meetings up, the disconnections and the relations not established, the exclusions. All this contributes to the specificity of place.

(Massey, 2005, p. 130)

However, the practical and methodological implications of relationality and specificity remain undisclosed in Massey's theoretical explication of space as the sphere of political potential and multiplicity (2009, 2005, p. 4). Massey's intention is never to pursue a precise solution or process by which the implications of political and spatial multiplicity might be stabilised and resolved; any such simplification would be counter to the embedded specificity and open-endedness of such practices. However, in not providing tangible and positive applications of multiplicity, Massey leaves a series of questions regarding the implications and potential of her positive re-articulation of space.

It is in this regard that our chapter and wider book posits Hamdi's development practice as a novel interpretation of space as a practical multiplicity, subsequently intersecting with our critique of how the projection of Western values that can accompany institutional development might be challenged by the concept of multiplicity (Fichter and Turner, 1972, p. 133; Hamdi and Goethert, 1989, pp. 23–24). Here Massey's interdisciplinary critique of Johannes Fabian's notion of coevalness becomes valuable, offering tangible practical methodological insights into multiplicity by appropriating ideas from post-modern anthropology and re-reading them against the negotiations of value and textual questions of space found in participatory development practice.

Coevalness

Reflecting many of the observations of Said's critique of *Orientalism*, Johannes Fabian notes the implications of placing 'those who are observed' in a different time from 'the Time of the observer' (2002, p. 25). This system of structural Western abstraction thus 'sanctioned an ideological process by which relations between the West and its Other, between anthropology and its object, were conceived not only as difference, but as distance in space and time' (2002, p. 147), thus 'time is used to create distance in contemporary anthropology' (2002, p. 25).

Considered in the context of our study the concept of coevalness articulates the explicit realisation that interactions of architects and development practitioners with 'other' communities can never be truly neutral. Consequently, any positive and open dialogue between communities and practitioners must be understood and practised on an even playing field; practised with and between partners of mutual respect and equality. This contestation of the coevalness of encounters with multiplicity leads Massey to the eventual summation that '[c]oevalness concerns a stance of recognition and respect in situations of mutual implication. It is an imaginative space of engagement: It speaks of an attitude' (2005, p. 69). It is this contestation of *the attitude* of engagement that provides a mechanism for positive, concrete, and critical comparison of multiplicity with the development practice methodologies of Nabeel Hamdi.

Crucially for this chapter, Fabian draws a distinction between the practical and theoretical aspects of anthropological discourse in critical observations of the 'temporal distancing' (2002, p. 82) that is implied, created, and reinforced by the denial of coevalness. This observation of coevalness as a relation of time provides an analytical critique of the geometries of power (Massey, 1991, pp. 25–26) at play in the dialogue between anthropology and its objects of study – placing them temporally in the past to observe them. It implicates the global and local alike in the problematic questions of interactions between different identities as representations of development, progress, and (in contemporary contexts) the ideological cohesion of neoliberal capitalism (Bauman, 2000; Morris, 2010, p. 110).

Fabian's analysis describes a great variety of 'distancing devices' each contributing towards a global result which he terms 'the denial of coevalness' (2002, p. 31). This observation of actual systemic devices of abstraction implicates anthropology in a far wider academic and political tendency to abstract and isolate others and otherness 'in a Time other than the present of the producer of anthropological discourse' (Fabian, 2002, p. 31). Thus Fabian's advocacy of a post-modern anthropology as a confrontation of coevalness as a reflexive social praxis provides an opportunity to re-evaluate the implications of communication, dialogue, and 'confrontation with the time of the Other', with the values and spatial relations found in places of difference and alterity: 'I also believe that the substance of a theory of coevalness, and certainly coevalness as praxis, will have to be the result of actual confrontation with the Time of the Other' (Fabian, 2002, p. 153).

By re-reading this 'distancing' as an aspect of both time and space, Massey's critical interdisciplinary comparison of the negative affects of anthropological and spatial abstraction resonates with our underlying exploration of dialectical materialism and spatial practices. What remains is to transcend the use of coevalness as a concept for the negative critique of space, and instead to articulate a potentially positive re-interpretation of coevalness as a spatial methodology with which to re-imagine Westernised architecture and space.

Massey's utilisation of Fabian's notion of coevalness is built upon a critique of his analysis of time as the pre-eminent distancing factor in anthropological discourse and practice. Instead, she observes that because time and space are interdependent then space is intrinsic to the notion of coevalness. This intrinsically connects coevalness to her critique of the fallacy of global hegemonic development as an inevitability by engaging with Fabian's advocacy of the primacy of mutually reflexive human productions of space in the relationality of localised practice and praxis (2005, p. 18, 1994, pp. 2–5). Taking precedent from Said's recognition of the epistemological distancing of Orientalism (Lewis, 2007, p. 67), Fabian's conception of the implications of time as a device for subordinating 'the Other' similarly becomes a methodology of introspection into the space and spatial practices of anthropology and ethnography: 'Through the distancing and objectifying depiction of a seemingly unaffected Other, anthropologists forgo a critical self-reflection that would render them a constitutive part of a hermeneutic (and thus "coeval") dialogue' (Bunzl, 2002, p. xiii).

In this critique of 'the distancing and objectifying depiction of a seemingly unaffected Other' this chapter suggests there is an explicit recognition of the interdependence of space, time, and language in Fabian's critique of anthropological praxis. Furthermore, the concept of 'coeval dialogue' is of particular interest in comparison with development practice given Fabian's recognition that the denial of coevalness is an overtly political act, suggesting further implications for the contestation of global development ideologies and practices, and the practical spatial methodologies employed by Hamdi in his pursuit of small change and sustainable social development.

Forward reasoning

Forward reasoning is a term used by Hamdi to describe the systemic perception of development that pervades practices by governments and NGOs across the informal (and formal) spaces of the Global South (2010, p. 155). In essence it describes a system of development that originates from centralised and structuralist models of government, leading to global, regional, and local policies made on assumptions from the top down (Hamdi, 2010, p. 12).

Based upon this hierarchical structure the analysis, projects, and solutions of such development are described by Hamdi as exemplifying a model of 'forward reasoning': in essence thinking from the top-down planning and abstract perspective without reference to the contingent reality of everyday life and the material reality of grass-roots practice. Instead, such development is based upon analysis drawn from global policy and national government issues, and quantifiable policies and politics (Burnell, 2012, p. 135; Hamdi, 2010, pp. 4–12; Hamdi and Goethert, 1989, p. 25). These solutions are then forcibly imposed on (or at best adapted to) localised conditions, leading to unsustainable, fragmented, and short-sighted development practices.

The power and authority that global institutions and NGOs possess are built upon inherently structural and hierarchal concepts of scale and influence (Rist, 2006, p. 6). The benefits and aspirations of such projects are found in the concept of impacting upon the most people by creating international policies and programmes of development. Such abstracted and quantifiable 'forward reasoning' planning values have inevitably become the institutional core of development policy (Elmore, 1979, p. 606).

This way of planning by forward reasoning is governed by global policy, risk assessment, ideological assumptions, and the structuralist measurement of value and achievement as objects and numbers instead of engaging with the everyday life and reality of real people (Bebbington, 2004, pp. 278–281; Brown, 2004, pp. 238, 249; Elmore, 1979; Hickey and Mohan, 2004, pp. 161–162). The specific objectives and steps designed to achieve such policy goals are projected upon local contexts along with quantifiable structures for measuring success.

Examples of the potential damage of this approach can be found throughout development discourse including the implications of oppressive houses observed by Turner in 1950s Peru (Fichter and Turner, 1972, p. 56), or Hamdi's various observations of the ill-conceived housing projects that overlook cultural conventions in Thawra (2010, p. 23), state immigrant camps with cartels controlling services with fear and violence (2010, pp. 45–49), and Betty the buffalo donated by an NGO (2010, p. 107). These are exemplars of a development process that is predominately conceived of in abstraction and isolation from the social and material reality of specificity and relational place.

Such spatial practices oppress, homogenise, and reject the rich multiplicity of space that organically emerges from informal and alternative communities of practice. Here we can instantly recognise similarities to the political and planning processes at play across Westernised society and space. Whether in the Global South or Global North, forward thinking provides development via institutionalised policy making, and does so without a clear sense of a coeval engagement with any form of material or dialectic foundational reality on which to support ambitions of socially sustainable development.

The broader implication of such largely well-intentioned yet abstracted development aid reflects precisely the impetus of Hamdi's contestation of forward reasoning (Esteva, 2010; Gronemeyer, 2010; Hamdi, 2004, p. 16). Hamdi observes that such examples remain a prevalent presence in contemporary development countries, observable in the prescriptive spatial realisations of Western values that are complicit with projecting aspirations of implausible and abstracted notions of how other people should live (2004, p. 11). This deterministic approach to development is realised in idealistic and ideological notions of space that are subsequently concretised and expressed in architectural forms and spatial relations; forms, spaces, and relations which exist in order to affirm the inevitability of a singular universal model of what being developed actually means (Hamdi, 2004, p. 13).

This is not to suggest that large scale development planning and systematic governance is inherently damaging and detrimental. Indeed Hamdi is quick to highlight that his discourse does not advocate a rejection of such systems (2010, p. 157), and would agree in principle with much of the work of Jennifer Brinkerhoff (2002; 2007) and Gaby Ramia (2003), for example, who continue to discuss and pursue the potential of global development policies and ideas such as the Millennium Development Goals (MDGs).

However, Hamdi advocates the necessity to challenge, balance, connect, and correct such forward reasoning with grass-roots observations and participatory practices and 'backward reasoning' (2010, p. 183). It is in this suggestion that Hamdi reveals an advocacy of what could be interpreted as coevalness in reaction to the distancing and abstraction of formal, centralised, and hierarchical planning by forward reasoning. In essence he advocates a reversal of the trajectory of projecting global ideas down, describing instead a reciprocal and dialectic process of scaling local ideas up, learning from them, then using planning systems to support the spread, adaptation, and growth of good ideas to foster wider regional and global connections. In doing so he is explicitly seeking to generate an interdependent dialogue and learning process between policy and the streets. This bottom-up model of spatial agency is an implicit engagement with the idea of space and development as a multiplicity. This is in direct challenge to forward reasoning's logic of coherence which is constantly renewed and reinforced by:

> the myth that practice can be controlled from the top, because that is where it starts, driven by experts whose business it is to ensure compliance with national and international norms and standards, agreed globally. It assumes that policymakers are adequately equipped or even well enough informed about the appropriateness of policy in the mess of practice. Its tendency is to assume normative standards of correctness of success. It is the logic and reasoning of providers, top down in bias, working often from the outside in.
>
> (Hamdi, 2010, p. 156)

Within Hamdi's observations is a profound contestation of the idealistic conviction of inevitable prosperity being realised through development so long as we all share 'similar objectives, values and beliefs' (2004, p. 13). Hamdi's critique of this model of ideological development reflects a practical awareness of the material, economic, and social reality of such informal places, and of the reality of coeval confrontations with difference and otherness. Here Hamdi's advocacy reflects an awareness of the disparity between abstract idealist and material subjective values at the raw and un-sanitised edges of development (Hamdi, 2010, p. 224; see also Jameson, 1991, p. 117).

As observed by Fabian and Massey, distancing and abstraction generate an inability to engage with the spatial relations, interactions, and material reality of such different and informal contexts. Ultimately this brings into question the appropriateness of the outcomes of development. Thus, Hamdi's observations provide rich evidence of the implications of approaching the practical complexity of the reality of the everyday and the values of the 'Other' through the fixity of an administrative, abstract, and academic monologue:

> The expert comes to be seen as a special kind of person, rather than that every person is a special kind of expert. Power relations are reinforced. All of which reflects in the behaviour and relationships to people who become beneficiaries rather than partners to our work.
>
> We wind up diagnosing people and their condition of poverty, as if it were some form of avoidable malignancy. [...] We contradict others who may not share our view of right or wrong, good or bad. We judge and stereotype those whose views and habits we find odd, but which may be entrenched in cultural norms and practices about which we may have, at best, a partial understanding. We will often label as troublemakers the loud or the pushy in a community and so exclude the very people who can get things done. And because we are experts, we wind up lecturing

rather than dialoguing. When dialogue becomes monologue, we seed the beginnings of all kinds of social injustice.

(Hamdi, 2010, p. 145)

Significantly, this distinction between dialogue and monologue can be understood as part of the practical methodological distinction between forward and backward reasoning. Resonating with Fabian's articulation of coevalness as a praxis (2002, p. 153), Hamdi's methodologies bring into question the physical, linguistic, and symbolic relationships that are played out within grass-roots participation. Building upon this, Hamdi is able to frame the implications of such interventionist prescriptions of value through the documentation of multiple subaltern experiences of forward reasoning development:

> In the end we got a building, a centre, he said. We went along with their ideas, nodded our way through endless meetings, talk shops and flip-chart presentations because, as always, getting something, we thought, is better than nothing and, besides, they had good intentions. It would have been impolite to question their wisdom and judgement, to challenge their authority. They were, after all, well educated. They had come a long way and were here to help.
>
> (Hamdi, 2004, p. 62)

Such experiences as this reflect the social, spatial, and material implications of forward reasoning as experienced at grass-roots level. The impacts on social identity, empowerment, and equality that structural and hierarchal development models are crystallised by a mis-conception and projection of values expressed through a spatial practice of monologue authority and social expediency. Such practices of forward reasoning inherently believe and rely upon the assumption that they can see the whole picture at the start. This logic assumes that end results can be formalised, planned, and executed as a means to achieve success and value as a universal assumption of what development means and represents (Burgess et al., 1997, p. 19; Fichter and Turner, 1972, p. 61). Yet the expectations of trickle-down economics from top-down planning are not only felt in their un-sustainable economic inefficiency but also in the continuation of development as a denial of coevalness:

> They had been treated as beneficiaries. It was, they said, a process without dignity, despite the generosity of donors. It lacked 'social intelligence' or caring. It was insulting and wasteful. It was all about charity and not about development.
>
> (Hamdi, 2010, p. 30)

In light of these observations, Hamdi's use of a disarmingly simplistic and apparently self-evident distinction between dialogue and monologue in development practice methodologies offers an incisive critique of the values that such practices suggest (2010, p. 145). It begins to contest the implications of both prescribed and negotiated values within post-colonial development. In this critique the monologue is thus symptomatic of structural and deterministic Western notions of 'modernisation' that preclude a singular inevitable constitution of what development means and the values it instils (Brandon, 1976; Esteva, 2010, p. 15). This is not to refute the potential of universal ideas such as democracy which have historically been the positive foundations of global development aid, but to bring into question the notion of Western democracy as being the only vision of how democracy is lived and enacted (Massey, 2005, pp. 66–70).

In the context of this comparison and taken to its full theoretical extent, forward reasoning as a monologue can be interpreted as synonymous with value prescriptions based upon the continued contemporary effects of globalised inequality (Lummis, 2010; Sassen, 2012). This re-reading of forward reasoning development provides a practical interdisciplinary exemplar of Fabian's anthropological denial and distancing of coevalness. It represents a denial of the material and social implications of the true proximity and engagement with the multiplicity, difference, and otherness of people existing at the informal peripheries of the world (Hamdi, 2010, pp. 164, 224, 2004, p. 118). In this comparative re-reading the cultural monologue of planning by forward reasoning highlights a social distancing of the true multiplicity necessitated by the coeval confrontation with difference and otherness, and what Massey would describe as the 'throwntogetherness' of space (2005, p. 141).

Such observations and comparisons are not intended as an explicitly derisory or retrospective criticism of genuinely well-intentioned aid-based development. Such actions must be appreciated and critiqued as laudable reactions to the pressures of a violently growing and urbanising population as discussed in the introduction and the statistical horror of 840 million people globally malnourished or 1.1 billion people without access to safe drinking water, and so on (Hall et al., 2013; Hamdi, 2010, p. 8; World Bank, 2013).

Instead the criticisms of top-down development founded upon forward reasoning are offered as a means to positively contextualise Hamdi's methodological distinction between forward and backward reasoning, and between singular or plural interpretation of values (Bhabha, 2004, p. 249; Hamdi, 2010, pp. 33, 54). Hamdi exposes the methodological and practical implications of forward reasoning as a monologue of broadly idealistic yet inherently prescriptive values, but crucially also diagnoses the pattern of such practice in the spatial, cultural, and practical inability of development practitioners to engage and interpret values through listening to the voices of others:

> Their compassion, she often found, was degrading not comforting, their good intentions made matters worse because they raised hopes and expectation in ways which could never be achieved.
>
> (Hamdi, 2010, p. 23)

And again similarly:

> Mistakes abounded. Some were technical, others reflecting ignorance, misunderstanding or disdain of culture and habit.
>
> (Hamdi, 2010, p. 25)

In comparison with Massey's and Fabian's observations and advocacy of coevalness and the implications of temporal and spatial distancing we are able to offer a theoretical reinforcement of Hamdi's critique of forward reasoning. The practical implications of multiplicity and coevalness in explicit connection with developing sustainable economic, social, and political values can now be interpreted as an explicit critique of forward reasoning and the structural hierarchies of global policy abstraction that is often only experienced by practitioners at grass-roots levels and felt by the 'known' receivers of such aid. As such, this remains a gravely inaccurate proposition for the sustainable development of the rising global informal population, however well intentioned they may or may not be (Fichter and Turner, 1972, p. 61; Neuwirth, 2012, see also 2006).

In being explicitly framed by the need to see space from different perspectives, the connection of multiplicity, spatial proximity, and dialogue with sustainable and coeval development practice now offers an opportunity to explore the theoretical implications and potential of Hamdi's 'backward reasoning'.

Spatial and textual distancing

Fabian's alternative notion of 'confronting the Other' poses a revelatory spatial question in relation to development. It offers a critique of the relation between the spatial, visual, and linguistic as a methodology for questioning the complex relations of inequality between the apparent 'knower' and the 'known' of development:

> From detaching concepts (abstraction) to overlaying interpretive schemes (imposition), from linking together (correlation) to matching (isomorphism) – a plethora of visually-spatially derived notions dominate a discourse founded on contemplative theories of knowledge. As we have seen, hegemony of the visual-spatial had its price which was, first, to detemporalise the process of knowledge and, second, to promote ideological temporisation of relations between the Knower and the Known.
>
> (Fabian, 2002, p. 61)

In this link drawn between the visual-spatial and the relationship between the interlocutors, actors, and agency ('the knower and the known') in any spatial practice, we find the theoretical suggestion of alternative spatial development practice as a coeval act (Bunzl, 2002, p. xxii). Building upon the potential of post-modern ethnographic field work as such a coeval spatial act, Fabian focuses on a critique of the disparity between such practices and their academic representation.

The critique observed by Fabian of 'distancing' and 'the denial of coevalness' in theoretical discourse is overtly proposed to be an expression of time as interdependent with development. Yet Massey's geographical and global application of this concept makes far more explicit the spatial contingency of 'distancing' (2005, p. 70). Her extension of Fabian's alternative interpretation and construction of anthropological practice towards an explicitly spatio-temporal critique provides insight into the challenges, implications, and political necessity of coevalness as a criterion for alternative spatial interactions. This notion of alternative and participatory spatial practice has been explored in Western architectural and spatial theory (Blundell Jones et al., 2005; Fraser, 2012; Hall, 2003; Till, 2009). Yet the alterity of such approaches can only be made explicit by the contestation of authority and control of meaning and intent that is implied by coevalness.

Using coevalness as a critical lens with which to reconsider the spatial interactions of Hamdi and development highlights the innovation of his shift from traditional models and methodologies of interaction with other people and places. This distinction and transition in his work and discourse impli-cates questions of authority, ideology, and value in explicit interdependence with the practical methodologies of engaging with openness, respect, and responsibility to those receiving development aid (Gronemeyer, 2010, p. 68). Furthermore, it also reveals the epistemological significance of what a coeval spatial and temporal condition of practice entails. Thus in a comparable comment, Bunzl notes the practical implications of Fabian's discourse in the realignment of the anthropological self and ethnographic other in the 'moment' of field-work itself:

> Praxis as an epistemological alternative to the allochronic rhetoric of vision [...] demands the conceptual extension of the notion of praxis to the ethnographic moment of fieldwork itself. In this sense, he not only propagates the critical textual reflection of fieldwork as an intersubjective – and thus inherently dialogical – activity, but paves the way to a coevally grounded conceptual realignment of anthropological Self and ethnographic Other.
>
> (Bunzl, 2002, p. xiv)

By pursuing a comparative analysis of these notions against recognised development practice methodologies we can offer some more concrete obser-vations on the real implications of such a politically provocative notion as coevalness as a practice. The implications of this comparison in the context of broader post-structural discourse is implied in Buntzl's implementation of the notion of 'textual production': 'Much like Orientalism, Time and the Other represented the synthesis of a politically progressive and radically reflexive epistemology with a critical analysis of the rhetorical elements of textual production' (Bunzl, 2002, p. xxii).

In a spatial context we must appropriate this concept of 'textual' as recognising that, as with texts and words, people read, interpret, and enunciate space in sub-jective ways, creating a multiplicity of individual relationships with space. The premise of considering development (or architectural and urban design more commonly in Westernised space) methodologies and practices as 'textual' reinforces an inherent self-reflective questioning of the presumptions and con-ditions of 'the outsider' and its interactions with 'otherness'. This self-criticality is most clearly contested through the power balance and inequality of com-munication in practice, where the intersubjective relations of spatial agency are played out through the authority and assumptions that are held within language and values. This interdependent relationship between presumptions, language, and values can be seen as a clear indictment of old-paradigm notions of colonialism and development (Hamdi, 2004, p. 118), but crucially also suggests the potential for alternative dialectical negotiations of difference, meaning, and values that define development itself and development practice:

> I advocated a turn to language and a conception of ethnographic objectivity as communicative, intersubjective objectivity. Perhaps I failed to make it clear that I wanted language and communication to be understood as a kind of praxis in which the Knower cannot claim ascendancy over the Known (nor, for that matter, one Knower over another). As I see it now, the anthropologist and his interlocutors only 'know' then they meet each other in one and the same contemporality. [...] If ascendancy – rising to a hierarchical position – is precluded, their relationships must be on the same plane: they will be frontal. Anthropology as the study of cultural difference can be productive only if difference is drawn into the arena of dialectical contradiction.
>
> (Fabian, 2002, p. 163)

The backward reasoning of buffalo mozzarella

As a means of elaborating on the comparison of coevalness and forward reasoning it is useful to explore an example drawn from Hamdi's participatory development practices and discourse. Wonderfully it affords an opportunity to discuss a project instigated by Hamdi that began with the observation of Betty's buffalo. By following an alternative backward reasoning methodology, Betty became the starting point from which Hamdi pursued development that was relationally specific to that place at that time, and which offered a scaleable model for socially sustainable development (Hamdi, 2010, p. 106).

Betty was named after a women working for an NGO that had previously been active in the village in which Hamdi found himself working. After find-ing themselves unable to dispense with a substantial amount of development money quickly and visibly, the buffalo project had been reasoned by this NGO as the most viable means to rebuild livelihoods; that is, forward reasoning (Hamdi, 2010, p. 107). This in itself acts as a representation of the mentality that can be fostered by 'forward reasoning' development, and encountered by both those genuinely seeking to engage in positive development and those seeking quick fix political gains.

> Sino and Tiba had nodded their way through endless monologues from Betty (the woman) on livelihoods, sustainability, self-realisation, cooperation and trust building, which the buffalo project was to inspire. They nodded more out of politeness than understanding. She was, after all, well intentioned and had come a long way. In any case, they had accepted the gift of the buffalo, as had others.
>
> (Hamdi, 2010, p. 108)

Whilst listening and talking to Sino and Tiba, Hamdi recognised that whilst there were great benefits of these buffalos gifted to the community, the NGO's projection of these animals upon individuals in a complex community had allowed their true potential to be overlooked. The potential to grow, breed, and take this opportunity to scale had been missed because the buffalo were being treated as isolated resources and objects for individual families and not for the potential of their communal value. Yet if the buffalo were bred carefully and utilised thoughtfully the interconnection of a community could be instigated around simple sustainable ideas to support the growth of small businesses – that is, local dairy production (not just individual subsistence farming) and the cottage industry needed to support it. Thus, Hamdi's approach is framed around these seemingly small and simplistic observations which are explored freely and organically without presumptions of values or outcomes:

> We had stumbled upon the beginnings of a narrative that would serve to discipline the design of their community facility. Later that day, we extend and enriched this narrative with other community groups, building on the aspirations of people and all their resources of talent and skill and speculating on outcome.
>
> (Hamdi, 2010, p. 109)

This opportunity had been recognised by listening to and learning from the material and practical realities of people and place (Burnell, 2012, pp. 137–138; Hamdi, 2010, p. 25; Hamdi and Goethert, 1989, p. 10). Building upon this initial observation using the same practices of dialogue, learning, and open-ended ambiguity of practice (2010, p. 156), Hamdi soon began to generate a viable programme for a purposeful and sustainable community-centre building:

> They would need a place to for making ceramic pots to pack the curd – a pottery, which might itself extend to making pots for other markets. There would be a place to weave and embroider cotton patches that are

typically used to cover the curd pots. Much of this activity would be home-based. The centre would offer opportunities to socialise around work and for training. Someone had the idea of turning buffalo dung into smaller 'mosquito coil' type pellets, easily scented with herbs and then marketed as organic mosquito repellent, crude but effective. Then there would be cheese-making, their own brand of mozzarella, their own label. There would be training in book keeping and marketing, offered through the Women's Bank, and later on a shop and cafe. This would be the start to a number of urban farms or enterprise centres nationwide. One would dream, in time, of a federation, a networked organisation joining the Fair Trade Alliance and competing for markets.

(Hamdi, 2010, p. 109)

Here forward reasoning is exemplified in the traditional approach of providing abstract things (in this case buffalo) without valuing the material and social reality of such complex and alternative communities. In contrast the same buffalo actually had the potential to be a scaleable catalyst for an inter-connected economic, social, and spatial enterprise (Hamdi, 2004, p. 12). Thus, whilst it soon became apparent that not everyone who has a buffalo and can benefit directly from this first idea, by applying the same backward and lateral reasoning to the social and material context Hamdi documents the creation of interconnected spatial relations that would validate and give sustainable purpose to a new community centre (2010, p. 118).

Hamdi makes clear that rather than having to create a community out of thin air and minimal resources, this process is simply looking to connect with existing enterprises and act as a catalyst of critical mass (2010, pp. 110–118). This would eventually see connections and projects with existing local enterprises and community groups including mushroom-growing cooperatives, local self-building programmes, recycling, education, and social buses: 'It would be like a laboratory, locally owned and managed and, in time, self-financing. It was all about partnerships, enterprise and livelihoods and importantly about building community and all kinds of assets' (Hamdi, 2010, p. 112).

Underpinning all of these connected projects was sustainable local finance (Hamdi, 2010, p. 42, 2004, p. 73) and social relations, and the dissemination of education and skills allowed the project to spread in a network of grass-roots initiatives that could not yet be perceived, but were waiting to be utilised by the entrepreneurial locals. By using a methodology of backward reasoning Hamdi had engaged with and learnt from the practical, material, and social reality of the settlement (2004, p. 140) and had allowed global agendas of governance and livelihood to be localised and made specific (Massey, 2005, p. 103). Reciprocally the interdependence and independence of the sustainable enterprise had allowed a scaleable project that had global potential for development change and resilience.

In the context of our study's comparisons, it is clear from this example that the contrast between backward and forward reasoning is a perceptual gap between the values and ideology of development as global policy initiatives, and the practical and material reality of the informal and subaltern communities and identities. Whilst the practical challenge of backward reasoning is highlighted in the development discourse that is critiqued by Hamdi and cited in this chapter, the theoretical implications of backward reasoning have previously been completely under-theorised or valued. Thus our comparisons seek to reveal the value of Hamdi's practices as exemplifying coevalness, open-ended ambiguity, and the negotiation of textual and hybrid socio-spatial relations and values that they imply (Massey, 2005, p. 85). Working from an abstracted outside perspective using abstract Western assumptions of value and development it would be impossible to perceive the potential challenges and opportunities of such a simple project.

In a direct context of neoliberal and Westernised space this sense of abstraction in the decisions that define development policy, architecture, and space is pervasive. Take for example UK planning policy for housing development which, despite apparent local council planning decision control, is extraordinarily biased towards supporting a model of standardised and speculative projects that are fundamentally damaging both socially and economically (Hatherley, 2011; Mace, 2013). The few examples of what we might describe as backward reasoning in Westernised space are usually exhaustively studied and explored by academics in far greater detail than the overwhelming majority of mediocrity that defines our collective social space.

Take for example 'Renew Newcastle' in New South Wales, Australia, founded and run by Marcus Westbury (2015). This joyous example of twenty-first century crowdfunding and grass-roots participatory change is profoundly encouraging yet astonishingly frustrating. It is now shown to be a vital example of how to revive and redevelop failing urban manufacturing and retail spaces, but has so far received very little academic engagement from planning discourse and development policy. The opportunity for 'Renew Newcastle' style development in post-industrial cities across Westernised space is tantalising specifically because it reflects so much of the backward reasoning development logic employed by Hamdi in his spatial practices.

Yet for now in Western space the pervasive assumptions that drive development remain those of abstraction, producing social spaces and relations that undermine the sustainability and efficiency of both global and local development. And by not engaging in textual and material dialectics of practice they also continue to project identities of development hegemony that Massey would decry as temporal convening and the homogenisation process built upon a fallacy that applies to both the Global North and Global South, namely the need for everyone to 'catch up to the West' (2005, pp. 82, 124).

Relational learning and backward reasoning

Our comparison to Fabian's theoretical discourse of 'coevalness as practice' (2002, p. 35) provides a critical foundation for a further interrogation of the implications for perceptions of value and meaning offered by alternative spatial practices like those explored by Hamdi. Building upon this, Hamdi's observation and practical valuing of other people's narratives and perspectives highlights the inherent potential of dialogue and learning (2010, p. 175, 1986) in his advocacy of development framed by the grass-roots knowledge exchange of textual places and backward reasoning (2004, p. 130):

> Backwards (reasoning) assumes essentially the opposite: the closer one is to the source of the problem, the greater is one's ability to influence it; and the problem-solving ability of complex systems depends not on hierarchical control but on maximising discretion at the point where the problem is most immediate.
>
> (Elmore, 1979, p. 606)

Returning briefly to a critical reflection on post-modern anthropology, perhaps the simplest practical interpretation of coevalness for Fabian begins with the idea of the self-aware practitioner (2002, pp. 60, 66; Gardner and Lewis, 1996, p. 113). This recognition of the impossibility of neutral and abstracted interactions with 'others' in anthropology and ethnography is countered instead by an advocacy of active engagement with the inherent political, economic, and cultural relations that are carried by actors and agency into development spaces (N. Long, 1992, p. 20). This recognition of the necessity of self-reflection as a means to critically discuss the asymmetric power inherent in the discourse and dialogue of practice again demonstrates a potentially valuable comparison with post- or alternative development practice as advocated by Gardner and Lewis:

> [post-modern] anthropology promotes an attitude and an outlook: a stance which encourages those working in development to listen to other people's stories, to pay attention to alternative points of view and to new ways of seeing and doing. This outlook continually questions generalised assumptions that we might draw from our own culture and seek to apply elsewhere, and calls attention to the varied alternatives that exist in other cultures. Such a perspective helps to highlight the richness and diversity of human existence as expressed through different languages, beliefs and other aspects of culture.
>
> (Gardner and Lewis, 1996, p. 167)

This awareness of the observer as actor in post-modern anthropology is recognised in contemporary notions of 'situated analysis' and 'standpoint theory' (ibid., pp. 22–23). Such approaches advocate both discourse and action as a field in which cultural and anthropological interpretations and interactions with 'otherness' are related to the observer's own subjectivity and political, cultural, and economic context. This relational self-awareness of actor-orientated spatial practice suggests a critical emphasis on the interaction of differences and values between actors and cultures, raising broad questions of how to define what development means and whose values it reflects (Chambers, 1997; Lummis, 2010; Sachs, 2010).

Building on Foucault's discourse on the plurality of knowledge (2001), Gardner and Lewis recognised that anthropological knowledge must always be understood as inherently political; thus 'the criteria of what constitutes knowledge, what is to be excluded, and who is qualified to know involves acts of power' (1996, p. 71). Here Said's critical historical analysis of ideologies being built upon assumed ontological stability and supremacy (2003, p. xii) can be seen to resonate with post-modern anthropological practice and the confrontation of coevalness. Subsequently the interdependence of political, ontological, and linguistic post-structural theory is only truly understood through socio-spatial practices that are coeval:

> The ontological presuppositions of the researcher are therefore not considered to be more complex than the ones ascribed to the local actors themselves. This means that the researcher does not occupy a privileged position; he or she can no longer choose between the attitude of the observer and a performative attitude, but places his or her own interpretations on the same level as the actions and expressions of the actors.
>
> (Seur, 1992, p. 139)

In this context, our comparative analysis once again suggests that this critical self-awareness is comparable to Hamdi's backward reasoning approach to development practice as a process of dialogue. In this comparison the importance of development as a spatial practice is reinforced by the recognition that such alternative practices cannot be reduced to the instigation of concrete architectural solutions as per top-down interventionist development. Instead they are explicitly framed as dialogues, processes, and practices that question, challenge, and produce values. By listening and learning from the social and material reality of places, Hamdi's methodologies are a dialogue with the economic, socio-cultural, and political values and the spatial relations that might be infused in such alternative spaces. Consequently, they inherently articulate evolving and continuous spatial practices of coevalness seeking to find and learn from the value and potential in the complexity of plural and relational identities:

> Patterns of social differentiation then are only made meaningful when situated in terms of everyday social practices and situations. In other words, it is necessary to show how relationships, resources and values are contextualised (actualised) through specific action in contexts, and the focus on action is central to the endeavor.
>
> (A. Long, 1992, p. 164)

Perhaps the clearest and most provocative analysis of the implications and potential of coevalness and backward reasoning is found in Andrew Long's advocacy of understanding such practices as 'where a joint construction of meaning takes place at the interface with outsiders' (Villarreal, 1992, p. 265). This intersection of development and anthropological theory represents an inversion of any surviving imperial sense of moral, political, or cultural authority; a radical reinterpretation of the traditional relational equality of development relationships. Long, Fabian, Massey, and Hamdi each in their own way describe practices of situated analysis and coevalness as a process of open, reflexive, and continuous learning through interaction. The intersection observed here proposes a reinterpretation of social and spatial context as a dynamic, interdependent and emergent interface: 'defining a situation (or appropriate context) is an achievement made by actors themselves. The definition of the situation emerges from the interaction itself, and cannot be given merely by the structure of a wider arena' (A. Long, 1992, p. 164).

The imperative of spatial proximity and coeval dialogue as inherent to Hamdi's methodologies of backward reasoning and participatory practice are thus critically comparable to post-modern anthropology's contestation of the equality of interface between differences. Thus, if Hamdi's development interventions are 'continuously being modified by the negotiations and strategies that emerge between the various parties involved' (Villarreal, 1992, p. 264), and confront a mutual process of value negotiation, what are the ethical and methodological implications for development practice? And what might such a critical inversion of the coeval agency of producing space suggest in reflection of Western spatial relations and practices?

A positive discussion of these issues is perhaps best understood in the recognition of the alternative, situated, and textual value offered by Hamdi's advocacies of 'listening to other people's stories' (2010, pp. 18, 62, 169, 174) and, as Gardner and Lewis highlight, 'paying attention to alternative points of view and to new ways of seeing and doing' (1996, p. 167). This attitude in itself suggests a great challenge to the conventional identity of both Westernised architecture and top-down development more generally.

In contrast with top-down forward reasoning Hamdi advocates methodologies that actually revel in and thrive upon the confrontation, interaction, and negotiation of differences and in the reality of inequality and economies of absence (2010, p. 12). He recognises that by actively engaging in this spatial proximity, the material reality of informal settlements becomes the very foundation for creating innovative, realistic spatial relations of sustainable development: 'many of the constraints we confront in the mess of practice are a context for work rather than a barrier to it' (Hamdi, 2010, p. 164).

In accepting and embracing the necessity, opportunity, and value of untidy answers, and open-ended practices, Hamdi's methodologies for observation, interaction, and facilitation of such traditionally unexplainable relationships become a vital point of critical analysis and validation for the theoretical contextualisation of his practice's explicit engagement with 'dialogue' that we have observed in this chapter (2010, p. 144, 2004, pp. 98–99). However, for Hamdi what is clear is that the connection between practices of spatial proximity and alternative forms of interaction between various interlocutors, actors, and agencies implies a fundamental challenge to the conventional identity and model of development practice. Here the profound theoretical, practical, and ideological implications of the distinction between forward and backward, monologue and dialogue are made clear:

> In all these respects, we are not good listeners because talking, not listening, is how you prove yourself – how you silence the opposition. It then follows, because we are not good listeners, we cannot be good learners – that sociable side of 'knowledge transfer' rather than 'knowledge hoarding'.
>
> (Hamdi, 2010, p. 12)

Simply put, if you allow yourself to get right in the deep end of all the mess and difference of other people you are confronted with alternative values and ways of living, alternative needs, and realities (Hamdi, 2004, pp. 4–6). The need to listen and be comfortable listening (Highmore, 2002, p. 177) to these differences is imperative for creating positive and meaningful dialogue (Hamdi, 2010, pp. 19–20). Hamdi repeatedly outlines, documents, and explains his methodologies of analysis in informal settlements through the very simple yet profound practice of listening (2010, p. 156, 2004, p. 58).

In this disarmingly simple advocacy of listening there is a provocative methodological insight and complexity that is easy to overlook. In the context of this chapter's premise the implications of Hamdi's recognition of the act of 'Listening, and importantly, being understood as one who wants to listen' (2010, p. 18), can be critically compared to Massey and Fabian as a contestation of the values of development that can only be achieved through a spatial practice of humility and coevalness.

The theoretical implications of practices that listen for and subsequently find opportunities for development is explicitly valuable in comparison with Massey's advocacy of the multiplicity of space being interdependent with the coevalness of encounters with difference. In exploring this idea of a re-valuation of the act of listening when engaging with differences and otherness, the practical reality and methodological implications are discussed by Hamdi in the interdependence he inscribes between practice and learning (2010, pp. 169–175, 2004, pp. 125–128, 1986), noting how 'Open learning is about cultivating mutual respect, about building each other's capacity to learn and influence practice – to be catalyst for change in each other's world, not just our own' (Hamdi, 2004, p. 128).

This practical realisation of dialogue as a reciprocal act of mutual learning and coevalness articulates a new positive notion of place-making as a situated material practice. Hamdi's practices offer a concrete realisation of coevalness, connecting practice to learning, mutual respect to creativity, negotiation to sustainability, and value to freedom: 'The mutual impingement of relations of power and difference within and across different arenas, conditions possibilities for agency and voice, as it does the value and purpose of learning' (Hamdi, 2010, p. 171).

It is important here to build upon the notion of learning as intrinsic to the idea of practice. Hamdi observes this distinction that in fieldwork learning is self-ordered, and that coeval practice is a process in which 'you learn what you need as you go along and do what you need to do to learn' (2004, p. 125). This notion of experimental, adaptive, and open learning practice connects with the importance of development as a dialectical practice; a spatial practice, but also a continuous practice. Thus, Hamdi observes: 'Change (however) only sticks when we understand why it happened. Continuous change is, therefore, contingent on progressive learning' (2010, p. 138), offering a startlingly simple realisation and alternative elaboration of Lefebvre's analysis of 'the social production of space', but crucially, doing so through a methodological criticality of intersubjective otherness, difference, and positive heterogenous multiplicity:

> In so doing, we enable people to find new ways of doing, thinking and relating in response to everyday problems which one takes for granted – breaking down barriers; optimising not maximising. These are all the qualities of leadership in practice and for development – a new openness for dialogue and learning.
>
> (Hamdi, 2004, p. xxiv)

Hamdi's backward reasoning and dialogues of coevalness evolve into the notion of open and continuous learning for both sides of encounters of difference. The connections between listening, respect, and learning link to an alternative vision and methodology for development practice that revels in the mess and contingency that coevalness requires. In so doing they are able to foster relationships of mutual respect and partnership; the inception of a notion that space is a social process, not a product.

Once again, in the context of Westernised space there are positive examples of backward reasoning engagements with space and values as a dialogue. In the twentieth century a notable example is the wildly under-appreciated work of Giancarlo de Carlo over several decades at Urbino in Italy (1970). His city masterplan and interventions in both new buildings and radical substantial renovations engaged with both the social and physical fabric of the city in a way that is rarely seen from Western architects. It is likely that the deep care and attention needed to achieve this success in masterplanning a city is connected with the years of engagement de Carlo has committed to Urbino. This dialogue, continuing over almost decades, provided not only a platform for de Carlo to truly understand the nature of this small city, but also necessitated a reciprocal responsibility for de Carlo to have a sustainable long-term plan for it.

At the opposite end of the spectrum of dialogue and backward reasoning in Westernised space we might also question the emergence of pop-up architecture in the twenty-first century (Beekmans and Boer, 2014; Oswalt et al., 2013). These developments in temporary and informal architecture have followed the path of all cultural trends, becoming commoditised and fetishised as objects at the expense of engaging with the real social, economic, and political potential such informal and adaptive architectures could empower (Bishop and Williams, 2012; Hou, 2010).

Spatial practices based on backward reasoning such as those outlined above can be seen to provide distinct concrete realisations of the inter-dependence and dialectical nature of political, economic, and socio-cultural spatial relations. Yet whilst these concrete achievements should not be over-looked, they also raise questions of negotiation, language, and value which in the context of our study remain the most provocative theoretical progression of Hamdi's observations:

> the need to achieve that base of interdependence 'when we no longer have to assert our individuality and independence against the world, because we are secure in ourselves and can achieve recognition of ourselves as separate, coupled simultaneously with our inevitable dependence on others'. In this way, we move from a position of 'us and them' to one of 'we'.
>
> (Hamdi, 2004, p. 138)

This observation of the necessary shift from the position of 'us and them' to one of 'we' is the focus of the theoretical foundations observed already, and the subject of the next steps of critical comparison in this chapter. Yet instead of the negative critical questions of identity and authority outlined by Said, further comparison of Hamdi with the key post-structural theories of Bhabha and Spivak will seek to suggest the positive potential of development con-frontation, negotiation, and production of textual value through listening and learning. Hamdi's observations will continue to provide the practical realisations of the theoretical critique of value and reality:

> All of this gets you involved, very often, in things you don't normally do or intend to do but have to, and other things you know you shouldn't do but do anyway to get jobs started. It gets you focused on pursuing ideals, not just project objectives.
>
> (Hamdi, 2010, p. 164)

The principles of backward reasoning offer a provocative practical and spatial engagement with the material reality of informal contexts. It invokes the necessity of practices that seek to engage in dialogues that reveal, contest, and negotiate the power relations of informal space in order to generate interdependence and mutual respect. As an alternative to the inevitable hierarchical power struc-ture of a policy-based and abstract reading of informal space, this chapter contests that such alternative development practices are based upon the social and political value of 'enunciation' (Bhabha, 1990, p. 312) and a 'negotiation of meaning' (Bhabha, 2004, p. 254). They are based on the realisation that only through active participation and negotiation derived from physical and interpretative proximity with places where problems exist can the value of development truly be explored.

Third spaces of open and reflexive learning

When critically compared to wider discussions of value in post-colonial theory and post-structural discourse, Hamdi's disarmingly simple observations of the social and spatial value of dialogue and negotiations (2004, p. 127) suggest provocative theoretical implications for methodological 'backward reasoning'. This observation, practice, and advocacy of dialogue and the negotiations of value in informal spaces offers an intriguing connection to Bhabha's observations that 'all cultural statements and systems are constructed in this contradictory and ambivalent space of enunciation' and the inherent creation of 'third spaces' of cultural hybridity in acts of negotiations and enunciation of meaning (2004, pp. 53–56).

For Bhabha, '[t]he importance of hybridity is not to be able to trace two original moments from which the third emerges, rather hybridity [...] is the "third space" which enables other positions to emerge' (Rutherford, 1990, p. 211). Third space, as introduced and developed by Bhabha, is interpreted here as articulating a space where the assumptions and inevitability of social relations and practices are challenged and subaltern identities are confronted, practised, and produced by identities outside the assumptions cultural hegemony.[1] In this way, 'third space' represents an enunciative site that encourages 'inclusion rather than exclusion' (Meredith, 1998, p. 3) through an interrogative negotiation of heterogeneous cultural forms that blur hegemonic boundaries of polarisation. In essence, it is a space of coevalness with the potential to continually (re)produce identities, spaces, and social relations through practices of dialogue and participation. Thus it can also be intriguingly compared to the logic and processes of dialectical materialism and the potential for a Lefebvrian engagement with the social production of space (Shields, 1991, pp. 50–58).

Bhabha places political significance on the act and space of enunciation as a counter to the structural and ideological necessity to control and authorise the physical and theoretical inscription and transcription of value to signs and signifiers (2004, pp. 2, 9). He notes: '[t]he wider significance of the postmodern condition lies in the awareness that the epistemological "limits" of those enunciative boundaries of a ethnocentric ideas are also the enunciative boundaries of a range of other dissonant, even dissident histories and voices' (2004, p. 6).

Resonating with our previous study of the positive practices of catalytic disruption, Bhabha's third spaces of negotiation, enunciation, and plurality can be perceived as being critically articulated by Hamdi's participatory practices. Dialogue, coevalness, and reflexive open learning in such spatial practices expose the abstractions of structuralism to the chaos and 'thrown-togetherness' of material subjectivity and the spatial trajectories of a truly post-colonial (and more broadly post-structural) context (Bhabha, 2004, p. 35). Such interactions generate a textual questioning, and answers to practical problems, explored from within the social, material, and quotidian reality, exist inherently within the multiplicity of informal space. The negotiations and cultural enunciations of third-space hybridity are a theoretical articulation of the spatial practices of listening and dialogue that Hamdi's methodologies provide. Both sides of this comparison revel in the irrevocable ambiguity of positive cultural difference and cultural definitions of value (Bhabha, 1990, p. 312).

It is however, necessary to briefly reflect upon concern surrounding such post-colonial theories of cultural hybridity. The most often cited critique is that it exists only as a 'first-world' discourse, and is subsequently rife with potential misrepresentations of subaltern voices. Here, Gayatri Spivak is known for her rejection of 'hegemonic nativist or reverse ethnocentric narrativisation' that accompanies any attempts to 'give speaking parts to the colonised' (Parry, 2006, p. 37). Her critique appropriates much of Foucault's post-structural analysis of the dissemination of truth as a constitution of power through accepted knowledge, and thus the 'types of discourse which [a society] accepts and makes function as true' (Rabinow, 1984, p. 73).

Thus for Spivak, the term 'epistemic violence' (Morton, 2004, p. 90) becomes applicable in this sense, where Western academia attempts to speak for/as the subaltern and in turn appropriates their voice, rendering them mute.[2] Hybridity can be and has been discredited in this way as exclusively for the 'new cosmopolitan elite' (Sayegh, 2008, p. 6) and rather less consistent with the reality of migrant diasporas. What is perhaps missing from Bhabha's theory of hybridisation therefore is the 'pedagogy of men engaged in the fight for their own liberation' (Friere, 1996, p. 39) and the potential of processes or practices of a critical consciousness that is illuminated by their own daily cultural, political, and material contexts.

Further critique is assigned to this 'hype of hybridity' (Morton, 2004, p. 90) and the enthusiastic acceptance of third space as a valid strategic alternative to the overwhelming dominance of hegemonic power structures. This supposed 'fetishisation' (Morton, 2004, p. 90) of post-colonial terminology has been critiqued as engendering such an array of translations and re-interpretations that it reduces any potential authenticity in its abstraction from the everyday. However, in response to these critiques, the comparison in this chapter focuses upon the theoretical proposition that meaning and value are forever in a state of being produced and re-produced, tested and negotiated.

We have observed that such propositions reflect a similar material dialectic to more practical negotiation of value described by Hamdi through the act of listening and open learning. This re-reading and comparison suggests a potential response to the need for a pedagogy of liberation and the importance of backward reasoning in order to facilitate Hamdi's methodologies of 'finding answers to questions you didn't ask' (2010, p. 64). In response to the reality and implications of subjective value, Hamdi provides a passionate advocacy and compelling observational evidence for a 'reflective practice' that allows for interaction and interdependence with valuable and ambiguous explorations of value:

Reflective practice qualifies or disqualifies the assumptions we make and the value we apply when defining problems, setting priorities or evaluating alternatives before we intervene. It tells us about the appropriateness of the norms and standards we apply and take for granted, the process we adopt, patterns of behaviour we assume to be current or acceptable, or otherwise, about our attitudes and judgement. Reflection nurtures wisdom and is a corrective to over-learning in schools.

(Hamdi, 2004, p. 135)

The methodological simplicity of Hamdi's approach deconstructs the professional abstractions of forward reasoning and hierarchical problem solving, and instead seeks to ground development practice in the ambiguity of working in real spaces, with real people and real problems (2004, p. 40). Thus whilst the language and implementation of enunciated values differ greatly from Bhabha to Hamdi, the underlying comparison remains compelling. Hamdi repeatedly documents and values the following sort of observations (2004, pp. 5–9), recognising that in such simple practical moments, the illusory simplicity of universal values is shattered:

Then pop into any one of the houses you will pass. Look at the priority that people attach to income rather than comfort. How much of the house will be devoted to home-based enterprise, how and where do people cook, eat and sleep?

(Hamdi, 2004, p. 5)

This type of material and cultural confrontation with the material reality of difference forms the context for the negotiation of meaning and values in the development of informal settlements. Practical and physical interaction within such contexts of difference provides a direct confrontation with the multiplicity of interpretations of value, be it living standards, needs, desires, or necessities (see: Illich, 2010; Rahnema, 2010; Rahnema and Bawtree, 1997, pp. 10–14). The significance of Hamdi's advocating the necessity of empathetic (Krznaric, 2014) spatial and cultural interrogation of such informal spaces is the acceptance of value as not being a universal object of knowledge, but as something that can only be achieved through a participatory and agonistic process (Gardner and Lewis, 1996, p. 71). Here Hamdi's methodologies and practices of negotiation and difference begin to suggest a legitimate critical comparison to Bhabha's theoretical concept of cultural hybridity and enunciation (2004, p. 38). The term enunciation is used in this context as the social differences articulated in the translation and re-articulation of meaning across cultural divides. Exploring this comparison further Bhabha discusses the ambiguous location of culture and seeks to articulate the implications of a distinction between cultural diversity and cultural difference in comparison with language and value signification:

> Cultural diversity is an epistemological object – culture as an object of empirical knowledge – whereas cultural difference is the process of the enunciation of culture as 'knowledgeable', authoritative, adequate to the construction of systems of cultural identification. If cultural diversity is a category of comparative ethics, aesthetics or ethnology, cultural difference is a process of signification through which statements of culture or on culture differentiate and authorise the production of fields of force, reference, applicability and capacity.
>
> (Bhabha, 2004, p. 49)

Here the advocacy of cultural difference resonate with Massey's multiplicity of space. Tracing a trajectory throughout this book from Lefebvre to Massey and beyond, the dialectical materialism of spatial multiplicity here intersects with Bhabha's cultural difference as a process in which cultural meaning is realised through the agonistic contestation of politics and values and through active practice and community participation (2004, p. 34). Hamdi's advocacy of methodological practices that engage, disrupt, and contest spaces through negotiation can be read as an evocation of the same politicisation of space that Bhabha confronts through the process of enunciation and cultural signification (2004, pp. 256, 264).

In comparison to the context of Westernised space we can see similar advocacies of increasing awareness of the political impact of neoliberal ideology on space, but as yet very little practicable or methodological engagement with how to counter the socio-spatial implications of laissez-faire attitudes to the rights of citizens to public space (Minton, 2012). Yet in Hamdi's practices we are offered the documented reality of such a provocative cultural process of negotiation in the participatory development of informal settlement communities. By looking past the humility of examples such as buffalos, mushrooms, and pickles, the implicit potential for achieving positive alternative spatial relations are revealed in the comparison of such spatial methodologies to the complex theoretical discourses explored in this research and critical analysis.

Thus Hamdi's use of disarmingly simplistic terms to discuss the practical realities and implications of coeval dialogue and practice as a reflexive learning process reveal practices of negotiation and cultural hybridity through the agency of coevalness, backward reasoning, and negotiations of third spaces in participatory development (2010, p. 18). The practical, performative, and anthropological observations of methodology compared in this chapter intersect with Bhabha's advocacy of cultural difference as a process of enunciation (2006, p. 156). Subsequently, Hamdi's discourse can now be observed as offering a remarkably practical translation of post-colonial ethics within development practice and suggests a pronounced renunciation of political, economic, or moral authority, replacing it with the far more tangible reality of engagements of mutual respect through sustainable development and practices of small change.

Enunciation and negotiation

Hamdi's participatory development practices and methodologies of coevalness, learning, and hybridity implicate the production of space as a complex social process. His conception of dialogue and negotiations of spatial and cultural relations begin to question and negate the presumptions and prescriptions that typically accompany the power-geometries and social relations inherent within development interventions (Gardner and Lewis, 1996, p. 16). His methodological propositions generate a socio-cultural space of enunciatory ontology, where different values and ideas are negotiated, inscribed, and must be continually re-inscribed in the practice and production of spatial relations and values that express the multiplicity of materially subjective trajectories (Hamdi, 2010, pp. 72, 179). Thus, in comparison with Hamdi, Bhabha's articulations of cultural hybridity as a dialogue and enunciation can here begin to be described as a process of engaging with and asking 'textual questions' in order to produce 'textual answers' (1990, p. 3) to cultural and spatial relations:

> It is that third space, though unrepresentable in itself, which constitutes the discursive conditions of enunciation that ensure that the meanings and symbols of culture have no primordial unity or fixity; that even the same signs can be appropriated, translated, rehistoricised and read anew.
>
> (Bhabha, 2004, p. 55)

The articulation of third spaces in Hamdi's participatory development methodologies is as both a physical space of translation and enunciation and also the notional socio-cultural space of value ambiguity (2010, p. 69). The mutual and coeval participation of listening and negotiation through dialogue is framed around the necessity of agonistic (Mouffe, 2013) contestation of both power relations and spatial organisations, generating a methodology in which the lack of fixity and predetermined outcomes of projects allows for the emergence of textual answers and values enunciated in social hybridity (Hamdi, 2004, p. 98). Such practices of enunciation and signification of value reinforce the comparison of Hamdi's backward reasoning methodologies to Bhabha's advocacy of cultural difference and suggest the provocative notion that to find true expressions of textual value we might look to the informal peripheries of space.

Hamdi's use of what we might now confidently call post-structural spatial methodologies implies his practices empower the inversion of professional and post-colonial inequality that accompanies notions of the design and resolution of space as the meaning or definition of development (Young, 1995, p. 186). Thus whilst in Hamdi the profound philosophical implications of these notions are not made explicit, the comparisons in this chapter suggest that his observations can be re-read as positive practical realisations of the inter-subjective contestation of post-structural values. These are explicit methodologies that not only reveal the need to approach spatial practice openly and without prescription, but also the need for reflection and reflexive responses to the inevitable incompleteness and failures such a process entails:

> Learning in action at first demands that we evaluate what we did, and with others. What went well and what did not go so well, to whom and why. It is a participatory learning process in which those to whom the impact of intervention is greatest have a dominant say about its value. From these assessments and narratives we draw lessons and discuss to whom the lessons apply. Importantly we reflect on what impact the lessons have on the way we may have to reorganize, or in the attitudes, tools, methods of practice, or on relationships between actors.
>
> (Hamdi, 2010, p. 171)

This built-in methodological self-criticism and open learning within such methodologies inscribes the interdependence and mutual relativity of all actors within such developmental practice (Long and Long, 1992, pp. 38–39). All interlocutors (internal or external, knower or known) are performing within a negotiated space and time – a third space – generating a process of auto-didactic learning and continuous value enunciation (Di Dio, 2012, p. 160). The broader political implications of such a process are made explicit when the enunciation of values is compared against the power relations inherent in the representation of space. Thus when the ambivalence of textual space is considered as an inevitable part of such processes:

> The textual process of political antagonism initiates a contradictory process of reading between the lines; the agent of the discourse becomes, in the same utterance, the inverted, projected object of the argument. [...] Reading [John Stuart] Mill against the grain, suggests that politics can only become representative, a truly public disclosure, through a splitting in the signification of the subject of representation; through an ambivalence at the point of the enunciation of a politics.
>
> (Bhabha, 2004, p. 35)

Re-contextualised against the idea of textual value, the inherent power relations and suggestive inequality of political representation and traditional development practices are inverted by the street level negotiations that Hamdi advocates (2010, pp. 9, 56, 87). Fracturing the ideological cohesion and structural abstraction of top-down development values, focus is placed upon active observation, interaction, and communication. Thus, contrary to first impressions of what appears to be simple humility and humble human interaction, Hamdi's practices suggest a re-valuation of development practice and communication, revealing the hidden opportunities and potentials of informal communities.

If so, Hamdi's practical methodologies of interaction, interrogation, and intervention within these spaces of development practice exist as a relatively unique documentation of negotiated and realised value and can be seen as a practicable counter-narrative to Western hegemonic formal architectural resolutions. In this light Hamdi's approach might be perceived as a withering critique of Westernised space. In this context we might broadly look at the social and economic implications of mass developer housing in the UK (Hatherley, 2009), or the privatisation of space (Minton, 2012). Further examples of socially divisive, ideologically conceived, and neoliberally value-engineered Western architectures might also variously include: Le Corbusier's housing at Pessac; Pruitt-Igoe by Minoru Yamasak; Torre David in Caracas; Westfield's Stratford East Olympic Park shopping centre; BDP's recent 'Liverpool One' city centre development.

Yet comparative analysis with the theoretical discourse trajectory of Massey, Fabian, and Bhabha suggests that Hamdi's practices and advocacy of backward reasoning must also be read as an implicit rejection of prescriptive and structural ontologies of meaning (Bhabha, 2004, p. 246). In contrast to the assumptions of Westernised space and development it advocates engagement with and learning from the material reality of informal contexts and the aspiration to offer a practicable realisation of textual value signification.

Learning, growth, discovery; the translation, signification, and enunciation of materialist and subjective values is here clearly defined as the fundamental principle of Hamdi's methodological approach. It requires an active and open participation, not only from the local population but also from outside practitioners seeking to advocate change and intervention. The inevitable dis-enfranchisement of prescribing and projecting the values of change upon the already charged context of informal settlements has to be overcome. Instead Hamdi notes:

> What we need, in this complex environment, is a kind of professional artistry which enables us to improvise and be informed, working somewhere between order and chaos, making what we can out of what we can get, making place without too much planning, making most of it up as we go along, a creative process of trial and error informed with experience and theory.
>
> (Hamdi, 2004, p. 116)

Thus backward reasoning is inherently more politically agonistic and yet potentially far less disruptive to the values and spaces that define informal settlements, and spaces that already exist on the peripheral edge of necessity, scarcity, and survival. And it is for precisely these reasons that the free and textual enunciation of values and identities becomes so fundamental to the process of development. The enunciation of space and value creates the opportunity for truly plausible development, but in doing so has to re-contextualise the inequality of power relations physically, economically, and culturally, as Bhabha notes:

> This emphasis on the disjunctive present of utterances enables the historian to get away from defining subaltern consciousness as binary, as

having positive or negative dimensions. It allows the articulation of sub-altern agency to emerge as relocation and reinscription. On the seizure of the sign, as I've argued, there is neither the dialectical sublation nor the empty signified: there is a contestation of the given symbols of authority that shift the terrains of antagonism. [...] This is the historical movement of hybridity as camouflage, as a contesting, antagonistic agency func-tioning in the time-lag of sign / symbol, which is a space in-between the rules of engagement.

(Bhabha, 2004, p. 277)

These processes of open and reflexive learning are inherently provocative to the broader notion of abstract professional expertise and suggest an inversion of the structural model of architecture, development, and space (Hamdi, 2010, pp. xvii, 16, 143). Yet if the object of development is to generate a plausible path for sustainable economic livelihoods these documented examples reinforce the necessity to fracture any prescriptions of value that arrive from Western contexts (Sachs, 2010). Instead Hamdi's methodological enunciation and communication of material specificity and reality suggests simply a pragmatic rejection of top-down, hierarchical, and abstract policy projected values for the pursuit of practicable and sustainable realities of multiplicity.

Textual value

For Bhabha, the connection between notions of subaltern identities and the necessary ambiguity of informal development space is played out in the con-testation of value signification (2004, p. 85). The hybridity and textual responses required to engage with the material reality and subjectivity of space occur through action. In re-contextualising Hamdi against post-colonial theory we are offered practicable, realised, and open-ended observations and methodological possibilities that further suggest the political possibilities and inevitable necessities of subaltern values in development practice. Once again, articulated with a disarming simplicity:

When we add the variable here and then for you, when we contextualise the question, it gives us a chance to ensure the answer itself is tailor-made to the specifics of place and people. The answer, in other words, will be different every time – it is open and even less certain.

(Hamdi, 2004, p. 131)

Uncertainty, and ambiguity, and the participatory empowerment of signifying your own values are perhaps thought of as obvious prerequisites of demo-cratic space. Yet when perceiving development using top-down and forward reasoning the ability to prescribe outcomes and abstract goals in advance of instigating development is inevitably considered to be the most efficient method to produce rapid intervention and solutions, whilst adhering to abstract, strategic, and structural models of space (Gardner and Lewis, 1996, p. 154). Yet the apparent logical cohesion of such development and spatial presumptions reveals the same continued potential for contemporary hegemonic ideology and authority to reproduce inequality. It is wonderfully rebuked in the above quotation in Hamdi's recognition of the massive implications of identity as a textually constructed context.

Both the desire to fix the world and the moral and ethical resolution to do so underpin the potential of Massey's positive spatial multiplicity; of socially and economically sustainable development, and Bhabha's enunciation of textual and hybrid values. Resonating throughout this book is the social necessity to define both what your own future looks like and the values with which it is produced. This is vital if space is to be realised to the full richness, diversity, and democratic pluralism of multiplicity. The comparative analysis in this chapter suggests that long-term strategic value is found not in abstract ideology, but in practices of uncertainty and the specificity of textual responses to spatial practice.

The political and post-colonial significance of our analysis of Hamdi's discourse is revealed in his attempts to navigate the narrative complexity of global and local power relations. His methodology of grass-roots practice using backward reasoning both confronts and contests socio-spatial inequalities in ways that seek to empower those who are not being heard or even have no voice (Hamdi, 2004, pp. 94–95, 110–115). Crucially, here we can see both the potential opportunities and the implications of backward reasoning through a comparison to Spivak's discussion of the inherent dangers of seeking to give voice to the subaltern:

> Subalternity is not that which could, if given a ventriloquist, speak the truth of its oppression or disclose the plenitude of its being. The hundreds of shelves of well-intentioned books claiming to speak for or give voice to the subaltern cannot ultimately escape the problem of translation in its full sense. Subalternity is less an identity than what we might call a predicament, but this is true in a very odd sense. For, in Spivak's definition, it is the structured place from which the capacity to access power is radically obstructed. To the extent that anyone escapes the muting of subalternity, she ceases to be a subaltern.
>
> (Morris, 2010, p. 8)

Based upon this notion of the fractured identity of those who might 'escape from subalternity', Hamdi's practices reveal a potential spatial methodology with which to navigate such an 'escape'. Practice, dialogue, and participation in the negotiation and enunciation of values in informal space explicitly seek to reveal such 'obstructions to access power' (Morris, 2010, p. 4) and through disruption and agonistic negotiation of value empower the subaltern to escape his or her 'political and cultural muting' (Morris, 2010, p. 104).

The comparison of Hamdi and subaltern theory confronts the fundamental political questions that exist at the core of interventionist development practice. However, our analysis of identity, values, and socio-spatial practice in the Global South also offers an implicit critique of the same political questions at the heart of architecture, space, and social relations in the Global North. The critical interplay of dialogue, empowerment, and political advocacy of 'others', and the equally complex 'right to speak for others' (Spivak, 1998) suggest the fundamental political implications of such contestations for people existing at the edges of informality, difference, and alternative space. Here the balance of coevalness and development practices that empower and facilitate local people's ability to contest their own space is confronted by a further theoretical intersection.

In her analysis of Spivak's discourse on identity Rosalind Morris observes that an inexorable question persists at the heart of post-colonial theory and subaltern identities: the critical question of whether it is possible or historically correct to empower[3] the subaltern to define his or her own freedom remains forever open and unresolved (Morris, 2010, p. 8). The danger remains for social relations that produce subaltern identities to become normalised and acceptable as part of existence and the timeline of development. Here Spivak's subaltern theory offers comparison with Massey's critique of passive space and global inevitability of cultural hegemony (2005, p. 110).

Thus, and notwithstanding the necessity to tread carefully amongst the endlessly complicated narratives of identity pertained to in subalterneity, there is potential in alternative development practice to see moments of comparative values. As we have observed, inherent in Hamdi's backward reasoning is the need for informal communities to be politically and culturally active as a means to empower themselves through agonistic yet socially sustainable spatial practices (Bhabha, 2004, p. 34). The more complex implications implied by subaltern theory are critically evidenced by observations of the socio-cultural implications of negotiation and enunciation for articulating mutual respect, humility, and coevalness in the confrontation of cultural difference as positive articulation of the multiplicity of space.

This comparison of Hamdi's discourse and practices offers a re-interpretation and much overlooked realisation of value negotiations and post-structural identities. Here negotiations based upon advocacy of subjective freedom of choice, autonomy, and control, are comparable against the materialist and textual realities of the structural inequalities of power relations within informal space:

> In this respect, the cultivation of choice when it comes to identity is one principal responsibility for all development practitioners, a central theme in participatory work, because the ability to choose, to adapt according to one's values, beliefs and aspirations, builds resilience and reduces vulnerability. It is a defence against having our identity coopted by systems, by planning ideals or single vision thinking. It builds resilience to exclusion and to violence.
>
> (Hamdi, 2010, p. 54)

The interdependence of learning and practice is advocated by Hamdi's metho-dological open-endedness, ambiguity, and unknowing as a positive and requisite implication of reflexive practice (2010, p. 156). This dialectical engagement with identity relieves the identity of authority and the control of knowing all the answers by creating the space for negotiation and learning. Backward reasoning provides the clearest and most succinct explication of such an approach, yet it is only by re-reading the simplicity of Hamdi against the complexity of Bhabha and Spivak that the implications of a post-structural comparison of such methodologies become clear:

> the 'negotiation' of the postcolonial position 'in terms of reversing, displacing and seizing the apparatus of value-coding', constituting a catachrestic space: words or concepts wrested from their proper meaning, 'a concept-metaphor without an adequate referent' that perverts its embedded context.
>
> (Bhabha, 2004, p. 263)

Hamdi's practical, cultural, and political distinction between forward and backward reasoning begins to connect to the global political implications of Spivak's conception of subalterneity and inequality. Similar deconstructive analysis can also be observed in the discourses of Fabian (2002), and Spivak (1990, pp. 225, 227, 228). For Spivak the question of value is necessarily complicated, loaded, and layered. She notes that 'if and when we ask and answer the question of value, there seems to be no alternative to declaring one's "interest" in the text of the production of value' (1985, p. 90). Yet as Massey observes (2005, p. 110) the true spatial and political value of Spivak's proposition is only ever experienced in the negative and as a counter-hegemonic narrative critique instead of a practicable opportunity (see also: Morris, 2010, pp. 88–89). In comparison with Hamdi, the analysis in this chapter offers a re-contextualisation of Spivak's post-colonial interrogations of value against concrete, realised, and documented socio-spatial methodologies of development. Thus the discussion of subaltern identity as a critique of authority and socio-cultural inequality can be brought to bear in comparisons to the development practice methodologies advocated by Hamdi and explored in this chapter.

Firstly, by advocating a backward reasoning approach to development that seeks a textualised response to socio-spatial relations and a negotiation of cultural hybridity and value. Secondly, Spivak's notion that 'revolutionary practice must remain persistent' (1985, p. 77) is suggestive of Hamdi's advocacy of grass-roots participatory practice (interpreted in the wider framework of our research as a model of dialectical materialism) as a means to generate social interdependence and change. This resonates with the suggestion that subjective and material value can only ever truly be constructed through an open-ended practice that is built open agonistic political process (Bhabha, 2004, p. 34). And thirdly Spivak's suggestion of the necessity of a textualised answer inevitably suggests the notion of a textualised question (1985, pp. 74, 82), and compares directly to Hamdi's methodological advocacy of human scale interaction and negotiation of value and the specificity of relations, necessities, and opportunities (Hamdi, 2004, p. 66).

The critical comparison and analysis offered in this chapter contends that Hamdi's methodology of backward reasoning can be perceived as a practical demonstration of approaching the production *and* practice of spatial relations *and* social value textually. Significantly it can be seen to offer a methodology for the practical negotiation of value multiplicity and a methodology that compares with Spivak's textual reading of linguistics and space (1985, pp. 88–90). Thus whilst attempting to maintain the pluralism and freedom suggested by the notion of subaltern identities, it is crucial to note how Spivak seeks to extract from the inevitable multiplicity of conjecture and ambiguity the clear political potential of materialist subjectivity. Provocatively, whilst intentionally rebuking the idea of pursuing a finite definition of subaltern values, Spivak arrives at the notion of materialist 'narratives of value formation' (1985, p. 82), providing a tantalising connection to the notion of subjective value negotiation through dialogue and participatory practice:

> The consideration of the textuality of value in Marx, predicated upon the subject as labor-power, does not answer the onto-phenomenological question 'What is Value?' although it gives us a sense of the complexity of the mechanics of evaluation and value-formation.
>
> (Spivak, 1985, p. 82)

In relinquishing the structural simplicity and abstraction of a monologue of Western universality Spivak is able to suggest that the question of post-colonial identity and value must be considered as necessitating a textualised answer, and what our critical analysis and comparison with Hamdi's development practices would describe as the multiplicity of textualised spatial practices:

> It is our task also to suggest that, however avant-gardist it may sound, in this uncovering, value is seen to escape the onto-phenomenological question […] if the subject has a 'materialist' predication, the question of value necessarily receives a textualised answer.
>
> (Spivak, 1985, p. 74)

In this context, Spivak advocates the ability to 'force' a post-structural textual reading of both universal and Marxist notions of value in order to contextualise a post-structural interpretation in the wake of the international division of labour (1984, p. 244). Further to this, Spivak's articulation 'that the moment of deconstruction of "philosophical" justice is the minute foothold of practice', crucially intersects with our comparisons between the deconstruction of universal values, meaning and identity and Spivak's aspiration to make Marx 'practicable' (1984, p. 244).

In this reflection on the interdependence of practice, identity, and social justice, Spivak's subjectivity of subaltern and post-colonial theory explicitly confronts issues of value and identity in an open-ended and ongoing dialectical process. Subaltern identities and values are crucial here because they are not able to be expressed without losing their inherent criticality. As such they reflect the same positive opportunities that Turner and Hamdi observe in the informal spaces and communities that are out of necessity outside and different to cultural hegemonic structures.

Yet the comparison of Hamdi's open-ended and reflexive practice to the inevitable intransigent and un-practicable qualities of subaltern discourse[4] remains open to further critical examination. The analysis in this chapter has provided a foundation of such a trajectory of critical interdisciplinary comparison, suggesting alternative practical contestations of the supposedly theoretical project of subaltern theories of identity and value in the methodologies of participation advocated by Hamdi. Here Hamdi's observations of the political necessity of open-endedness and ambiguity can potentially begin to be interpreted and critiqued as an (unconscious) attempt to navigate the practical reality and complexity of Spivak's various propositions for a textualised dialogue with subalterneity.

Developing values and textual learning from others

Our analysis in this chapter has sought a trajectory of interdisciplinary comparison that connects and contests Hamdi's development practice methodologies with a far broader and more complex range of spatial and cultural discourse. The underlying methodology built upon the intimacy and integrity of spatial proximity and coevalness reveals Hamdi's discourse and development work as a viable practical comparison to complicated spatial theory critiques. In these comparisons against the post-structural spatial theory of Massey et al., such alternative development practices based upon dialogue instead of monologue, or backward instead of forward reasoning must subsequently be re-read as generating alternative spaces and values through negotiation, enunciation, and signification (Villarreal, 1992, p. 251).

The critical exploration of the practical and theoretical implications of coevalness and dialogue generates a subsequent re-contextualisation of Hamdi's methodologies against the discourses of Said, Massey, Derrida, and Fabian. The comparisons observed here allow grass-roots participatory development practice methodologies to be re-considered as a method of negotiating the multiplicity of post-colonial spaces and of value translation and subjective inscription. These are spaces that offer the potential to transcend the problematic prescriptions of empirical, patriarchal, or hierarchical participation and suggest a truly reflexive, discursive, and reflective form of practice.

Such practicable methodologies are a clear engagement with the complex multiplicity of post-colonial space and the potential for the participatory negotiation of meaning and enunciated trajectories of alternative development.

Each of these interdisciplinary comparisons of coevalness, embedded material practices, enunciation, and subalterneity are drawn against examples of community development in the unexpected and unlikely examples of development practice as continued and evolving practices:

> planned intervention cannot be adequately comprehended in terms of a model based upon step-by-step linear or cyclical progression. Rather, it must be seen for what it is – an ongoing, socially constructed and negotiated process with unintended consequences and side effects. Applying this insight to the understanding of development projects and the differential responses they provoke, requires the deconstruction of orthodox views of policy and planning and of their capacity for steering change. We need alternative, more open and less presumptuous (hence less 'totalising') ways of thinking and acting.
>
> (Long and Long, 1992, p. 270)

Hamdi's methodological simplicity belies a far more profound expression of the potential of people and space to generate textualised answers to the reality of life at the turbulent periphery of development. The critical analysis in this chapter offers and further contests these methodological insights in relation to key post-colonial concepts outlined by Bhabha and Spivak. This subsequent re-interpretation of informal space and development practices resonates with the textual notions of value, advocating the negotiation and enunciation of social values (and thus relations) as perhaps the greatest chance to critically engage with the true multiplicity of a post-colonial world.

The critical analysis posed throughout this book is punctuated throughout with examples of practitioners and social actors working towards such engagements with the social production of space, relations, values, and identities, from within Westernised space. To say that such confrontations and contestations are only possible in the political and economic absence of the informal Global South is no longer a viable excuse as to why Westernised space continues to assume the inevitability of the social values and relations that accompany neoliberalism and the globalised inequality of capitalism. The simplicity of Hamdi's practices – backward reasoning, dialogue, and listening to others – is disarming. Yet the opportunity they present for Western architects, planners, politicians, and communities to re-imagine their space can only become increasingly pressing as the challenges of scarcity and inequality continue to press against our assumptions of capitalist development being able to deliver the fundamental fallacy of unending growth and equality.

As articulated in the comparisons proposed in this chapter, the achievement of Hamdi's spatial methodologies suggests an inversion of the assumptions of Western development inevitability, freedom, and value. The reflected implications of this upon Western articulations of architecture, development, and socio-spatial relations are left implicit within this analysis and open to further research and ongoing speculative questioning.

Alternative propositions for Western (or more generally, global) space founded upon networks of social actors, grass-roots self-governance, and development agency propose a methodological excavation of the complexity of community constructions of value. The implications of such approaches are the promotion of a never ending engagement and community discussion with their own economic, political, and cultural relations (Massey, 2005, p. 103). Yet it is discussion that provides the potential for a sustainable future, not any individual intervention, but discussion as a means to spatially engage with what development could be. The potential of such a critical empowerment of Western space would be to challenge and contest the notion of a Western authority, abstract space, and economic ideology as self-referential justifications with which to 'solve the problems of Others':

> Such a view does not eliminate an impetus to forward movement, but it does enrich it with a recognition that that movement be itself produced through attention to configurations; it is out of them that new hetero-geneities, and new configurations, will be conjured. [...] It is a politics which pays attention to the fact that entities and identities (be they places, or political constituencies, or mountains) are collectively produced through practices which form relations; and it is on those practices and relations that politics must be focused. But this also means insisting on space as the sphere of relations, of contemporaneous multiplicity, and as always under construction.
>
> (Massey, 2005, p. 187)

Notes

1 Here it is valuable to note the connection to Edward Soja's (1996) articulation of hybridity and third space as a projection of a Lefebvrian interrogation of space. Whilst this connection is intriguing it remains outside the remit of this book and is a subject for future research.
2 It is noteworthy that Spivak explicitly includes herself in this critique.
3 To empower or give power inherently implies control and authority in and of itself.
4 Qualities that are necessary in order to maintain the subjective implications of uncovering, interpreting, and discussing the subaltern other (Spivak, 1998, p. 84).

References

Bauman, Z., 2000. *Globalization: The Human Consequences*. Columbia University Press, New York.

Bebbington, A., 2004. Theorising Participation and Institutional Change: Ethnography and Political Economy, in: Hickey, S., Mohan, G. (Eds), *Participation: From Tyranny to Transformation*. Zed Books, London.

Beekmans, J., de Boer, J., 2014. *Pop-Up City: City-Making in a Fluid World*. Bis Publishers, Amsterdam, Netherlands.

Bhabha, H.K., 1990. *Nation and Narration*. Routledge, London.

Bhabha, H.K., 2004. *The Location of Culture*. Routledge, London.

Bhabha, H.K., 2006. Cultural Diversity and Cultural Differences, in: Ashcroft, B., Griffiths, G., Tiffin, H. (Eds), *The Post-Colonial Studies Reader*. Routledge, New York.

Bishop, P., Williams, L., 2012. *The Temporary City*. Routledge, London; New York.

Blundell Jones, P., Petrescu, D., Till, J. (Eds), 2005. *Architecture and Participation*. Spon Press, London.

Brandon, D., 1976. *Zen in the Art of Helping*. Routledge and Kegan Paul, London.

Brinkerhoff, J.M., 2002. *Partnerships for International Development: Rhetoric or Results?*Lynne Rienner Publishers, Boulder, CO.

Brinkerhoff, J.M., Smith, S.C., Teegen, H., 2007. *NGOs and the Millennium Development Goals: Citizen Action to Reduce Poverty*. Palgrave Macmillan, Basingstoke.

Brown, D., 2004. Participation in Poverty Reduction Strategies: Democracy Strengthened or Democracy Undermined?, in: Hickey, S., Mohan, G. (Eds), *Participation: From Tyranny to Transformation*. Zed Books, London.

Bunzl, M., 2002. Foreword: Synthesis of a Critical Anthropology, in: Fabian, J., *Time and the Other: How Anthropology Makes Its Object*. Columbia University Press, New York.

Burgess, R., Carmona, M., Kolstee, T., 1997. Contemporary Spatial Strategies and Urban Policies in Developing Countries: A Critical Review, in: Burgess, R., Carmona, M., Kolstee, T. (Eds), *The Challenge of Sustainable Cities*. Zed Books, London.

Burnell, J., 2012. Small Change: Understanding Cultural Action as a Resource for Unlocking Assets and Building Resilience in Communities. *Community Development Journal* 48, 134–150.

Chambers, R., 1997. *Whose Reality Counts?: Putting The First Last*. ITDG Publishing, Bradford, UK.

de Carlo, G., 1970. *Urbino: The History of a City and Plans for Its Development*. MIT Press, Cambridge, MA.

Di Dio, D., 2012. The Placemaker's Guide to Building Community – Book Review. *Community Development Journal* 47, 159–161.

Elmore, R.F., 1979. Mapping: Backward and Implementation Policy Decisions. *Political Science Quarterly* 94(4), 601–616.

Esteva, G., 2010. Development, in: Sachs, W., *The Development Dictionary*. Zed Books, London.

Fabian, J., 2002. *Time and the Other*. Columbia University Press, New York.

Fichter, R., Turner, J.F. (Eds), 1972. *Freedom to Build*. Macmillan, New York.

Foucault, M., 2001. *The Order of Things: Archaeology of the Human Sciences*. Routledge Classics, London.

Fraser, M., 2012. The Future is Unwritten: Global Culture, Identity and Economy. *Architectural Design* 82, 60–65.

Friere, P., 1996. *Pedagogy of the Oppressed*, 2nd Edition. Penguin, London.

Gardner, K., Lewis, D., 1996. *Anthropology, Development and the Post-Modern Challenge*. Pluto Press, London.

Gronemeyer, M., 2010. Helping, in: Sachs, W., *The Development Dictionary*. Zed Books, London.

Hall, S., 2003. Globalization from Below, in: Ings, R. (Ed.), *Connecting Flights: New Cultures of the Diaspora*. Arts Council/British Council, London.

Hall, S., Massey, D., Rustin, M., 2013. After Neoliberalism: Analysing the Present. *Soundings* 53 (April), 8–22.

Hamdi, N., 1986. Training and Education: Inventing a Programme and Getting it to Work. *Habitat International* 10, 131–140.

Hamdi, N., 2004. *Small Change*. Earthscan, London.

Hamdi, N., 2010. *The Placemaker's Guide to Building Community*. Earthscan, London.

Hamdi, N., Goethert, R., 1989. The Support Paradigm for Housing and its Impact on Practice: The Case in Sri Lanka. *Habitat International* 13, 19–28.

Hatherley, O., 2009. *Militant Modernism*. Zero Books, New York.

Hatherley, O., 2011. *A Guide to the New Ruins of Great Britain*, Reprinted Edition. Verso Books, London.

Hickey, S., Mohan, G., 2004. Relocating Participation Within a Radical Politics of Development: Insights from Political Practice, in: Hickey, S., Mohan, G. (Eds), *Participation: From Tyranny to Transformation*. Zed Books, London.

Highmore, B., 2002. *Everyday Life and Cultural Theory*. Routledge, London.

Hou, J. (Ed.), 2010. *Insurgent Public Space: Guerrilla Urbanism and the Remaking of Contemporary Cities*. Routledge, New York.

Illich, I., 2010. Needs, in: Sachs, W., *The Development Dictionary*. Zed Books, London.

Jameson, F., 1991. *Postmodernism, or, The Cultural Logic of Late Capitalism*. Verso, London.

Krznaric, R., 2014. *Empathy: A Handbook for Revolution*. Rider, London.

Lewis, H.S., 2007. The Influence of Edward Said and Orientalism on Anthropology, or: Can the Anthropologist Speak? *Israel Affairs* 13(4), 774–785.

Long, A., 1992. Goods, Knowledge and Beer; The Methodological Significance of Situational Analysis and Discourse, in: *Battlefields of Knowledge*. Routledge, London.

Long, N., 1992. From Paradise Lost to Paradigm Regained?; The Case for an Actor-Oriented Sociology of Development, in: *Battlefields of Knowledge*. Routledge, London.

Long, N., Long, A. (Eds), 1992. *Battlefields of Knowledge: The Interlocking of Theory and Practice in Social Research and Development*. Routledge, London.

Lummis, C.D., 2010. Equality, in: Sachs, W., *The Development Dictionary*. Zed Books, London.

Mace, A., 2013. Housing-led Urban Regeneration: Place, Planning, and Politics, in: Leary, M.E., McCarthy, J. (Eds), *Routledge Companion to Urban Regeneration*. Routledge, New York.

Massey, D., 1991. A Global Sense of Place. *Marxism Today* 38, 24–29.

Massey, D., 1994. *Space, Place and Gender*. Polity Press, Cambridge.

Massey, D., 2004. Geographies of Responsibility. *Geografiska Annaler Series B: Human Geography* 86, 5–18.

Massey, D., 2005. *For Space*. Sage Publications, London.

Massey, D., 2009. Concepts of Space and Power in Theory and in Political Practice. *Documents d'Anàlisi Geogràfica* 55, 15–26.

Massey, D., Featherstone, D., Painter, J., 2013. Stories So Far: A Conversation with Doreen Massey, in: Featherstone, D., Painter, J. (Eds), *Spatial Politics: Essays for Doreen Massey*. Wiley-Blackwell, Chichester.

Meredith, P., 1998. Hybridity in the Third Space: Rethinking Bio-cultural Politics. Presented at the Te Oru Rangahau Maori Research and Development Conference, University of Waikato.

Minton, A., 2012. *Ground Control: Fear and Happiness in the Twenty-First-Century City.* Penguin, London.

Morris, R.C., 2010. *Can the Subaltern Speak?: Reflections on the History of an Idea.* Columbia University Press, New York.

Morton, S., 2004. *Gayatri Chakravorty Spivak.* Routledge, London.

Mouffe, C., 2013. *Agonistics: Thinking The World Politically.* Verso, London.

Neuwirth, R., 2006. *Shadow Cities: A Billion Squatters, A New Urban World*, New Edition. Routledge, New York.

Neuwirth, R., 2012. *Stealth of Nations: The Global Rise of the Informal Economy*, Reprinted Edition. Anchor Books, New York.

Oswalt, P., Overmeyer, K., Misselwitz, P., 2013. *Urban Catalyst: The Power of Temporary Use.* DOM Publishers, Berlin.

Parry, B., 2006. Problems in Current Theories of Colonial Discourse, in: Ashcroft, B., Griffiths, G., Tiffin, H. (Eds), *The Post-Colonial Studies Reader.* Routledge, New York.

Rabinow, M. (Ed.), 1984. *The Foucault Reader.* Pantheon, New York.

Rahnema, M., 2010. Poverty, in: Sachs, W., *The Development Dictionary.* Zed Books, London.

Rahnema, M., Bawtree, V. (Eds), 1997. *The Post-Development Reader.* Zed Books, London.

Ramia, G., 2003. INGOs and the Importance of Strategic Management. *Global Social Policy* 3(1), 79–101.

Rist, G., 2006. *The History of Development*, 3rd Edition. Zed Books, London.

Rutherford, J., 1990. The Third Space: Interview with Homi Bhabha, in: Rutherford, J. (Ed.), *Identity, Community, Culture, Difference.* Lawrence and Wishart, London.

Sachs, W., 2010. One World, in: Sachs, W., *The Development Dictionary.* Zed Books, London.

Said, E., 2003. *Orientalism.* Penguin Classics, London.

Sassen, S., 2012. Inequality? We Need a New Word. *The Occupied Wall Street Journal,* 2 October. Available at http://occupiedmedia.us/2012/02/inequality-we-need-a-new-word/ (accessed 16 March 2016).

Sayegh, P., 2008. Cultural Hybridity and Modern Binaries: Overcoming the Opposition Between Identity and Otherness? Presented at the Cultures in Transit Conference, Liverpool University.

Seur, H., 1992. The Engagement of Researcher and Local Actors in the Construction of Case Studies and Research Themes, in: Long, N., Long, A. (Eds), *Battlefields of Knowledge.* Routledge, London.

Shields, R., 1991. *Places on the Margin: Alternative Geographies of Modernity.* International Library of Sociology, Routledge, London; New York.

Soja, E.W., 1996. *Thirdspace: Journeys to Los Angeles and Other Real-and-imagined Places.* Wiley-Blackwell, Oxford.

Spivak, G.C., 1984. Marx after Derrida, in: Cain, W. (Ed.), *Spivak, Philosophical Approaches to Literature: New Essays on Nineteenth and Twentieth Century Texts.* Bucknell University Press, Cranbury, NJ.

Spivak, G.C., 1985. Scattered Speculations on the Question of Value. *Diacritics* 15(4), Marx after Derrida, 73–95.

Spivak, G.C., 1990. Postcoloniality and Value, in: Collier, P., Gaya Ryan, H. (Eds), *Literary Theory Today.* Polity Press, Cambridge.

Spivak, G.C., 1998. Can the Subaltern Speak?, in: Grossberg, L., Nelson, C. (Eds), *Marxism and the Interpretation of Culture.* Macmillan Education, Basingstoke.

Till, J., 2009. *Architecture Depends.* MIT Press, Cambridge, MA.

Till, J., 2012. From Objects of Austerity to Processes of Scarcity. Available at http://www.jeremytill.net/read/98/from-objects-of-austerity-to-processes-of-scarcity (accessed 16 March 2016).

Villarreal, M., 1992. The Poverty of Practice, in: Long, N., Long, A. (Eds), *Battlefields of Knowledge.* Routledge, London.

Westbury, M., 2015. *Creating Cities.* Niche Press, Melbourne, Australia.

World Bank, 2013. *Inclusion Matters: The Foundation for Shared Prosperity.* World Bank, Washington, DC.

Young, R., 1995. *Colonial Desire: Hybridity in Theory, Culture and Race.* Routledge, London.

6 Architecture and space re-imagined?

The analysis and comparisons in this book have sought to build upon our original contention that aspects of Western critical theory and socio-spatial discourse can be valuably and provocatively compared against the practical realisations of pro-poor participatory development practitioners working in the Global South.

Re-reading and re-contextualising examples drawn from development practice has revealed spatial practices that deliver sustainable social enterprise by explicitly challenging the conventional approach and perspectives of Westernised architecture and development. Crucially, we have seen that such examples drawn from social, political, and economic contexts of the Global South reflect and resonate with key critical perspectives and theoretical aspirations of Western spatial theory.

Throughout these comparisons we have explored aspects of Henri Lefebvre's and Doreen Massey's urban and spatial theory, conducting a close textual reading of texts from their respective discourses. This approach has provided new perspectives and analysis of post-Marxist urban space, and an exploration of the explicit connections between Lefebvre and Massey in terms of the social production and multiplicity of space. This analysis generated a theoretical framework from which to reinterpret and revalue the approaches to participatory development practice found in the writings and projects of John Turner and Nabeel Hamdi. This research process provides a new method with which to re-read and critique Western socio-spatial theory. Subsequently, we arrive now at the questions of what can be learnt from contextualising the positive theoretical implications of alternative spatial practices of the Global South in order to implicitly speculate on their potential appropriation to the Global North.

The positive achievements observed in these examples of participatory development practice can thus begin to be seen to provide an implicit theoretical critique of Western spatial practices and conventional architecture. Such examples provide a rich new vein of alternative socio-spatial practices and examples from which to contest the seeming inevitability of Westernised space.

The original four cornerstones of this premise – Turner, Lefebvre, Hamdi, and Massey – have provided the foundations for an underlying critique of structuralist approaches and interpretations of space. In exploring these key protagonists, various unforeseen research trajectories have emerged. These thematic connections have provided opportunities to explore and critique a broader socio-cultural and political discourse. The connections uncovered range from agonistic political theory and post-modern anthropology, through to post-colonial and subaltern studies discourses. Yet ultimately each strand of research has sought to retain a line of critical comparison drawn between abstract theoretical discourse and concrete spatial practices, and the positive social values inherent within this re-imagining of architecture and space.

In this final chapter we will seek to summarise the critical observations, connections, and analysis offered in the preceding chapters, before attempting to provide synthesis and reflection as a means to frame ideas for further research and debate. Primarily we will attempt to use the observations and discussions raised in this book to re-visit the original intention of this research: to re-imagine Western space and architecture by learning from development practice in the Global South.

Materialism, choice, and autogestion

In chapter one we introduced and contextualised the premise that the development practice of Turner could be compared to the works of Lefebvre. Analysing examples from Turner's housing practices generated a critical lens through which to reveal and interrogate connections between disparate practical and theoretical discourses, and thus create a reciprocal re-contextualisation of both discourses.

Re-reading Turner in comparison with Lefebvre reveals how the principles and values that underpin housing development practices in informal settlements from 1960s Latin America can be re-valued. When considered in comparison with Lefebvre's observations of dialectical materialism and the social relations of production, Turner's observations demonstrate the economic and social value of progressive grass-roots development in producing and empowering meaningful housing communities.

The theoretical methodology of dialectical materialism can be re-imagined as exemplified by Turner's housing development in contexts of economic impoverishment. Turner utilises a Lefebvrian turn of space: inverting the assumptions of development, and challenging top-down dogma by engaging with the grass-roots material reality of informal space. Instead of development as a product, Turner advocated socio-spatial praxis to develop sustainable communities. Thus, Turner's principles of housing suggest a concrete realisation of Lefebvrian dialectical materialism.

Lefebvre's spatial appropriation of dialectical materialism is exemplified by Turner's practical methodologies for generating alternative spaces and social relations. Turner's provocative notions of 'user choice' and 'progressive development' can thus begin to be understood as examples of what Marx and Lefebvre would recognise as dialectical materialism, but explored in the context of economic and material absence. Placed in this critical comparison we can begin to re-imagine an alternative vision of Lefebvre's spatial appropriation of Marxist theory, namely, the concrete exemplars of alternative spatial practices realised in Turner's development practices.

The implications of this suggest Turner's development methodologies in informal settlements are an unexplored and un-critiqued realisation of Lefebvre's advocacy of a re-imagining of the politics of space and the social implications of the relations of production. Whilst these observations are specifically aimed at informal settlements, their explicit realisations of Lefebvre's positive aspirations for materialist and dialectic approach to space also suggest they exemplify something missing from contemporary Westernised space. The economic and social efficiency of Turner's user-defined housing provides a critical lens through which to consider the disjunction of use-values and exchange-values offered by informal and formal models.

Here our analysis returns reflectively to the canon on alternative spatial practices observable in Westernised space. What becomes clear is the lack of traction gained by Western protagonists of ideas like alternative housing; such as Colin Ward, John Habraken, Giancarlo de Carlo, Ralph Erskine, and even Nabeel Hamdi's work with the GLC. Notwithstanding various alternative housing models that have been pioneered in certain Western contexts (notably Holland and wider Scandinavia, where the political and economic models have afforded some successful largely middle class attempts at alternatives), these examples pale in comparison to the vast majority of debt fuelled housing that has dominated the past decades of Western housing models and continues to prevail in spite of the economy crisis of the 'sub-prime' housing markets.

Considered against conventional Western architecture and spatial practices, Turner's practices suggest a means to critically reflect and re-imagine the political and economic relations that define Westernised architecture and space. Yet these differences are articulated and experienced in the differing political and economic contexts of Global North and South. Whilst there are notable reasons why user-defined housing works (or is made to work) in the context of pro-poor development, it is also important to note the economic, political, and social impediments that would suggest it to be impossible to implement such practices in the West. Yet against this critical comparative lens of Turner's practice and Lefebvre's theory, the structural and quantifiable housing models offered in large-scale corporate and neoliberal Westernised cities and suburbs must also be critiqued as representations of an ill-conceived faith in the economic models of neoliberal capitalism (Hall et al., 2013, pp. 8–11; Harvey, 2005, p. 71).

Building upon the initial comparison, chapter one went further and critically questioned whether Turner's development practices reflect a post-structural re-interpretation of authority, identity, and values by engaging in grass-roots community participation. This comparison was further reinforced by the intersection with the theoretical discourse of Lefebvre's autogestion and self-management. Turner's progressive housing and community development offers a practical realisation of the social and political implications Lefebvre advocates through autogestion. This intersection and comparison of autogestion and grass-roots participatory practice thus provides a foundation methodology for alternative spatial practice and agency; practices explicitly built upon the logic of dialectical materialism.

The implications of this for our original premise of re-imagining Westernised space and architecture is significant. The necessity to re-read and re-evaluate Turner's work implicates a need to examine and contest the further potential of user-choice, autonomy, progressive development, and participatory practices as positive socio-spatial alternatives beyond the Global South. In connection with the material and dialectical logic of Lefebvre's aspirations for autogestion and positive socially produced space, participatory practices can be re-read as exemplifying the political and social potential of alternative spatial agency and architecture.

In re-reading Turner we can discern the necessity of user-choice and freedom in developing a socially and economically sustainable model of progressive housing and development. In comparison with Lefebvre, this analysis articulates the importance of understanding space both for its material reality and as an ongoing process. Echoing Turner's call to perceive 'housing as a verb', the potential concurrent re-imagining and re-articulation of (Westernised) architecture as a verb implicates people and space as interconnected in a continuous ongoing process that is integrated in the social and material reality of the everyday. It re-imagines architecture as a social agency directed towards the self-management and autogestion of space through sustainable social relations and practices.

Space and multiplicity

Chapter two provided an opportunity to connect and compare Lefebvre's spatial and urban theory to the contemporary socio-spatial discourse of Doreen Massey. Previously Massey's spatial interpretation of Marxism was considered to have emerged from her reading of Althusser (Featherstone and Painter, 2013, p. 4), however the research and analysis in this book provides an alternative perspective: specifically, the observation that both Lefebvre and Massey build upon a Marxist spatial turn to advocate the positive political potential of space as a medium for social relations of production, difference, and multiplicity.

The positive articulations of space that resonate in aspirations of both Lefebvre and Massey are fundamentally built upon Marxist and socialist conceptions of political space and the deeper fundamental logic of dialectical reasoning. This trajectory of analysis utilised Lefebvre's articulation of differential space as a projection of 'the right to the city' and connected to the spatial differences implied in Massey's conception of relational space as a multiplicity. Such comparisons revealed a wider constellation of connections to the political works of David Harvey, Chantal Mouffe, and Ernesto Laclau, as well as links to the provocative participatory development discourses of Andrea Cornwall et al. Ultimately this re-reading of connections between Lefebvre's concept of differential space and Massey's advocacy of the multiplicity of space provides potentially valuable contributions to the ongoing re-contextualisation of Lefebvre's ideals in a global and post-colonial context (Goonewardena et al., 2008).

Echoing Lefebvre's spatial methodologies and analysis, Massey offers a rich critical lens through which to perceive the structural limitations of interpreting space as mere representation of time and change. Massey's critiques of this 'taming of the spatial' (2005) provides an alternative interpretation of the interdependence of space and time as co-existing in the relational construction of societies. This new proposition suggests a continuity with the spatial aspirations advocated by Lefebvre, whilst also allowing for its contextualisation and grounding within a contemporary global context of inequality and geometries of power. The critical examination of structuralism's spatial fetishisation provides a further foundational meta-narrative concerning the positive critical comparison of the alternative development and socio-spatial practices in the Global South.

For Massey time and space cannot exist as dichotomy but must be understood as parts of the continually evolving dialectic process of the construction of social, political, and economic relations and values. The spatial concepts of appropriation and differential space exemplify Lefebvre's advocacy of the positive political potential of spontaneity and everyday life to transcend oppression and hegemonic space. Similarly, Massey's multiplicity is the recognition of other and alternative interpretations of the world as part of the relations that exist within space (and time). As the comparisons in this book have revealed, Lefebvre's and Massey's engagements with space, difference, and power are not best exemplified in Western space, but in development practices that critically engage with social, political, and economic contexts that confront conditions of global inequality.

Thus, whilst for Lefebvre space is social and emergent and real, for Massey space is coeval, relational, specific. The intersection of these conceptions of positive space and difference provided this research with comparisons to the development practices of Turner and Hamdi, but also expanded its context into further theoretical trajectories of post-colonialism and subalterneity.

These critical intersections of Lefebvre's and Massey's spatial advocacies provided the foundation for a post-colonial and globalised contestation of Westernised models of development, space, and thus architecture. The alternative social relations of space revealed by development practices in the Global South re-imagine space as a practical means to confront and challenge the global inequality and power-geometries of post-colonial development. This is pivotal in that it validates the comparison of development practices and informal settlements against Western spatial theories of the right to the city, to difference, and to multiplicity.

With this new contextualisation and comparison, the articulation of space and positive multiplicity in participatory development practice is implicated as a potentially invaluable new strand of critical spatial agency to be explored in Western discourse. This trajectory re-imagines Lefebvre's and Massey's spatial aspirations in a new global and post-colonial context, and crucially, not in abstract theoretical isolation but in participation and grass-roots social practices.

The significance of spatial difference and multiplicity for the re-imagining of architecture and space as a verb cannot be underestimated. To conceive of space as a practice and architecture as an agency of change requires engaging with multiplicity, difference, and pluralism as integral to the viability of culturally and politically active space. The confrontation of such open and positive difference challenges conventional Westernised architecture and space with issues of uncertainty and humility that are predominately cleansed from a profession built on certainty and authority. Re-imagining architecture as a verb is a rejection of space as inevitability and homogeneity, and thus the power that defining space as such implies to the architect and authors of space. This inversion remains a challenge confronted by only a few spatial innovators who recognise architecture as a positive advocate of difference, multiplicity, and open social change.

Geometries of power, spatial disruption, and scale

Chapter three explored the challenges and contradictions of key spatial themes in development practice: participation and hierarchy, authority and choice, practice and product. Building on Massey's analysis of Mouffe and Laclau, the concepts of hegemony and geometries of power expanded our comparisons with contemporary development practice. These renewed inter-disciplinary intersections contested Hamdi as an exemplar of counter-hegemonic spatial practices and participatory methodologies, specifically his use of practices of disruption, social catalysis, and 'small change' to deliver sustainable socio-spatial enterprise and development.

In reinforcing these comparisons, we explored a theoretical trajectory connecting Gramsci's interpretation of cultural hegemony with Mouffe and Laclau's advocacy of positive political spaces of agonism, before intersecting with Massey's critical contextualisation of the power-geometries of space. Here Massey's discourse is vital. It connects these concepts to a geographical and spatial critique of the inevitability of development under the political influence of capitalism and the economic implications of neoliberalism. And this critique of inevitability further reinforced our analysis of an explicit comparison to the alternative development practices advocated by Hamdi in informal settlements of the Global South.

This critical reflection and comparison of Hamdi's practices reveals how his methodologies reflect theoretical principles such as social agonism, disruption, and catalysis in order to produce positive alternative space. In this comparison, Massey's political and social implications of the relational specificity of space are explored by Hamdi in the agency and implications of his methodologies;

in practicable everyday contestations of post-colonial and globalised spaces of multiplicity.

Subsequently Hamdi's practices can be re-read as implicitly designed to reveal existing hegemonies and power-geometries that (re)produce the social and spatial relations of informal community. His alternative methodologies of grass-roots participation explicitly confront, contest, and agonise existing geometries of power in order to reveal the potential for catalytic projects of social and economic change. Whilst Hamdi employs these practices in contexts at the periphery of economic instability, in the context of comparisons with Massey et al., the social and political disruptions he generates are realisations of dialectical social change through the interrogation, disruption, and production of alternative sustainable social relationships.

Subsequently we revealed further comparisons to Laclau and Mouffe's contestation of capitalist ideologies of 'hegemonic logical cohesion' (2001, p. 3) and Massey's inevitability of neoliberal social relations of Westernised space. Hamdi's practices of 'small change' reveal and challenge spatial hegemonies, and in doing so, create the opportunity to empower and provoke change and alternative social practices and relations. The implications of such practical relations of positive counter-hegemony to Massey's post-structural discourse of space prompts a re-reading and re-contextualisation of her discourse as a theoretical framework within which to actively contest the practice and social relations of space.

In the context of this theoretical comparison Hamdi's explicit engagement in spatial methodologies that seek sustainable growth and transition from the small social and political disruptions provided a crucial link to the political and social relationality Massey conceives in the interdependence of global and local space. Subsequently, Hamdi's notion of the 'scaleability' of a social project or practice is thus observed in this research as crucial in combating the perennial problem of losing the necessary social and economic momentum that truly sustainable social change requires (Lefebvre, 1969, p. 84).

Here the social sustainability of alternative development is perceived in Hamdi's practices not as a rejection of capitalist economics, but a re-alignment of the purpose of capital and a confrontation, contestation, and diversification of the social relations that capitalism produces. This analysis suggests a clear and distinct comparative connection with the similar economic and political engagements we observed in Turner's alternative housing models in Peru. These practices are not a rejection of growth or a call for a socialist revolution, but instead are a contestation of inevitability and a re-politicalisation of the social relations and practice that produce space.

The introduction of Gramsci's hegemony and Laclau and Mouffe's agonistic space as political foundations of Massey's critique reframes participatory development practice as positive realisations of counter-hegemonic social relations. The economic and political disruption, contestation, and scaleability of Hamdi's practices suggest an engagement with political and social change that must be re-read in the context of Marx's, Lefebvre's, and Massey's critical spatial discourses. Notions of conflict and the contestation of space are crucial elements in the spatial practices of Hamdi in informal settlements, and clearly reflect the political and spatial need and necessity to challenge prevailing ideologies in order to see socially sustainable relations of production and change.

The importance of connecting concrete practices of disruption and small change to the cultural hegemony of Westernised space are integral to our overall aspiration to both re-imagine space and re-articulate architecture as a verb. Instead of limiting the potential for change, the comparisons explored here revealed the inherent instability of space and its openness to positive change and counter-hegemony. Learning from the humble small change practices of Hamdi contends an articulation of architecture as a verb and practice with the potential to contest hegemony and in doing so can reveal space as the medium for spatial agency and social change. By engaging in such alternative spatial agency, responsible architects can regain the same positive ethical agenda seen in Hamdi's development practices.

Considered in the context of global inequality and development the implications of specificity and relationality re-frame Massey's positive conceptions of space as a critical lens and theoretical framework in which to review the challenges facing counter hegemonic practices in the Global North and increasingly prevalent abstract Westernised spaces of neoliberal capitalism. Yet it also provided the basis for further research explorations of how identity is integral to the generation of positive social patterns and social relations, and especially those that might engage in positive counter hegemonies.

Identity and practice

Pursuing questions of authority, identity, and practice offered opportunities to compare and contextualise aspects of critical Western spatial theory to the historical and critical trajectory of development practice from Turner to Hamdi. This analysis intersected with important discourses and historical influences on the evolution of development practice in comparison with theoretical discussions of post-colonial identity and values. These critical comparisons observed the methodological evolution of development practice from Turner to Hamdi as mirroring several key notions from post-colonial and post-development theory. This opportunity to connect Turner and Hamdi with our wider underlying aspiration of re-imagining space and architecture was invaluable.

Firstly, the comparisons in this chapter offered a re-reading of the material and contextual practicalities examined in the work of Turner and Hamdi against the theoretical context of Edward Said's philological and historical contestation of identity, authority, and colonialism. This analysis successfully framed the comparison that these alternative approaches to space, identity, and development are also revelatory in comparison with the predominant forms of centralised and hierarchical development. This disjunction is overtly marked in the contrast between the formal centrality that Said theoretically observed in colonialism, and Turner's and Hamdi's contrasting engagement with informal and grass-roots practices. In contrast to conventional hegemonic projections and impositions of identity, Turner and Hamdi offer invaluable concrete realisations of space and identity as practices which engage with the positive political potential of difference, informality, and choice.

Secondly, the anlaysis in this chapter contested the implications of political identification and distinction between the developed and the developing worlds through critical comparisons with contemporary post-development discourse. This comparison was critically observed in connection with Massey's critique of the inevitability of development and space, allowing a critical contestation of Westernised identity as the pinnacle aspiration of development. Thus, the implications of the Global South and informal settlements as being 'under-developed' relates to Massey's contestation of spatial convening and the necessity of other cultures, places and identities to 'catch up to the West' (2005, p. 124).

Alternative contemporary post-development discourse is thus positively compared at the intersection of these comparisons with the participatory practices of both Turner and Hamdi. Our analysis positively contested such practices as exemplars of attempts to sublimate the restrictive implications of development identity as a product and mere reflection of Western and capitalist ideologies. Thus, this critical comparison and contestation of identity as a practice offers a re-reading of participatory and open-ended practice that seeks to define positive multiplicities of space and identity through social and community participation in the politics and practice of space, practices specifically advocated and embodied in the works of Turner and Hamdi.

The notion of identity as interdependent with the practice and production of social space once again provides provocative reflection on the state of con-temporary Westernised space. Taking the premise that space is thus a reflection of interdependent socio-political, economic, and cultural identities, this would appear to express a rather apt yet reflective critique of contemporary Westernised public and private space. The notion that identity is a practice implies a con-temporary return to Lefebvre's conception of 'space as a social product'. Yet in still seeking the full positive potential of a comparison with the development methodologies of Hamdi, our analysis of these exemplars of socially and eco-nomically sustainable spatial relations and practices reframes the potential of Western spatial practice towards possible re-imagination of architecture and space as alive, fluid, messy; an open, critical, and reflective social process.

By interrogating these practices against key post-colonial and post-development theory this analysis provides a re-reading and re-contextualisation of the social capacities and necessities of development practice as comparable to Lefebvre's social production of space and Massey's relationality and multiplicity of space, inequality, and global spatial relations. The question of identity as a product or practice implicitly intersects with the professional identity of architects in the Global North. Identity as a practice is equally as valuable as a means to help frame and articulate space and architecture as a verb in the Global North. The notion of engaging with and learning from the public, clients, and the people architects serve as equals is a fundamental imperative learnt from post-development and post-colonial theory. It provides the framework and exploration of value as a post-structural idea that is necessary to inform our final analysis: the intersections of development, space, and architecture as processes of textual and coeval practice.

Textual value(s)

Building upon this post-structural analysis of development practice, in chapter five we pursued further interdisciplinary comparisons of Hamdi's methodologies of practice as contestations of post-colonial identity and values. This analysis revealed theoretical and practical connections to the work of post-modern anthropology as well as the cultural theory of Bhabha and Spivak. The critical comparisons appropriated and leveraged theory from the post-modern anthropological advocacy of ethnographic spatial praxis of coevalness, mutuality, and equality. These comparisons revealed intersections with Massey's conceptions of multiplicity and difference of *other* communities. Her advocacy of the positivity, equality, and relationality of space as interdependent with time was thus able to be contested as an intersection with Fabian's pioneering advocacy of coevalness.

The analysis in chapter five provided critical comparison and analysis of Hamdi's development methodologies as exemplars of Fabian's principles of coevalness and the similar post-modern anthropological notions of 'situated analysis' and 'embedded spatial practices'. In this context, Hamdi's notion of engaging with people and space without prescribing the values or end results to his practices is a post-structural contestation of the necessary open-ended socio-spatial practices that development without authority and ideology entails.

Subsequently, this analysis framed examples of Hamdi's practices of dialogue and negotiation in comparison to the discourse of Homi K. Bhabha's notions of 'enunciation of meaning' and the hybridity of cultural 'third-space', and later to the deconstruction of values and authority proposed by Gayatri Spivak. These critical comparisons contested a re-reading of Hamdi's methodologies as politically, socially, and anthropologically nuanced articulations of sustainable social change, realised as an expression of values and identity.

This connection to Bhabha's concepts of enunciation and third space provides a contextualisation of both Hamdi's practices and Massey's spatial advocacy as potentially interdependent engagements with positive negotiations of cultural hybridity and difference. In this context Hamdi's methodologies of listening and reflective learning as participatory practice can be contested as exemplars of the emergence of textual identity as interdependent with the social production of space. Thus, the open-ended and coeval nature of Hamdi's practices were critically compared to a post-structural undecidability of meaning, and Spivak's advocacies of the textual value of otherness (1985).

This multi-threaded advocacy of the inherent instability of space – Massey, Bhabha, Spivak – allowed for direct comparison to Hamdi's methodologies of open-ended practice and of informal communities being engaged in the practices of defining their own meaning and values, and articulating their own (potentially alternative concept of) development. The necessary challenge to the excepted conceptions of the spatial expertise of architectural and development practice further highlights Hamdi's critical advocacy of the necessity of 'backward reasoning' as integral to the contextual material engagement with process.

The comparisons in chapter five provided perhaps the most speculative contestation of development practices against the post-structural theories of third space, textual value, and otherness. As such the situated practices of listening, learning, and backward reasoning in informal settlements as practices with which to engage and generate textual and subjective values are considered as exemplary post-structural spatial practice of previously unrecognised theoretical importance. Ultimately, Hamdi's (and by extension Turner's) practices must subsequently be re-read as offering a unique contestation and critique of the inability of hierarchical, formal, and conventional Westernised spatial practices to contest and explore values above mere formal and economic hegemonies.

This final critical comparison of Hamdi's work to the social and participatory enunciation of values supports the premise that sustainable social and economic development provides a fitting end to the trajectory of this book. The emphasis placed upon the concept of textual value perhaps best exemplifies the many themes explored, examined, and critically compared in our research. The contestation of meaning and textual values in space and practice provides perhaps the most provocative and challenging final reflected comparison to conventional Westernised architectural development and neoliberal spatial relations. These critical comparisons begin to suggest a positive articulation of architectural agency and spatial practice that implicitly and explicitly frames a discussion of textual values as interdependent with the social relations and spatial practices that produce space.

Considered in the context of a re-imagining of space and architecture, notions of coevalness, enunciation, and textual values can be considered as fundamental concerns of any socially responsible and sustainable spatial practice in both the Global South and Global North. The need to engage with people – be they clients, the public, developers, planners, or politicians – in free and open spaces of discussion defines the foundation of a positive articulation of space and architectural agency. Subsequently re-imagining Western architecture as an enunciatory and textual social process inevitably re-frames the profession itself as part of a newly open discursive landscape of coeval practice. By understanding the theoretical implications of enunciation and textual value to questions of social identity, and seeing in Hamdi methodologies with which to responsibly engage in such spaces with self-awareness, we can begin to articulate a plausible framework from which to positively re-imagine space and architecture.

These propositions are based upon a research trajectory that draws layers of practical and theoretical observations into critical comparison, generating a logical path of reason between previously disparate discourses. This interdisciplinary resonance between practice and theory provides a critical framework within which to conceive positive alternative social spatial practices of development as realisations of the counter-hegemonic spatial critiques of Lefebvre and Massey et al. Thus, advocating spatial practices founded upon listening and learning, the negotiation and enunciation of meaning, or engaging with informed and textual values, is not a rejection of the importance of architects and development practitioners. Quite the contrary. It provides a wide new framework of critical political and social engagement and empowerment to disillusioned communities and individuals who still pursue positive alternative spaces and social relations.

Questions and opportunities for further discussion

The analysis in this book sought to provide an exploration and examination of relationships, connections, and thematic resonances between examples of development practice methodologies and aspects of critical spatial discourse. Discovered using a methodology of close comparative reading these connections have validated our original premise of exploring the alternative economic, political, and social contexts of the Global South in comparison with key aspects of Western spatial theory. Highlighting and examining such thematic connections and resonances has provided new links between explicit issues of spatial theory and practice, the Global North and Global South, formal and informal, top-down and grass-roots socio-spatial practices. In the context of this research, alternative spatial relations and practices from informal settlements and peripheral space can now be perceived, valued, and utilised as practical realisations of key critiques and aspirations of Western spatial theory.

The analysis of these relationships has provided a wider framework of critical discourse and thematic exploration within which to value these examples within interconnected spatial disciplines. The examples explored in support of these comparisons reveal concrete realisations of key aspects of critical spatial theory, and the practical methodologies with which to begin to frame the wider project of contending assumptions of the inevitability of Westernised space. In light of this analysis, examples of alternative socio-spatial practice drawn from global economic peripheries begin to provide a framework from which to explore the critiques of neoliberal capitalism and Western ideology articulated by Lefebvre, Massey, and others.

This research provides a framework and entry point from which to explore this critique of Western spatial/architectural practice. More specifically it provides a methodology of comparison which can be used to examine the opportunities to learn reciprocally from development practice and Western spatial theory. This methodology of re-reading development practices from the Global South against Western spatial theory remains a valuable mechanism from which to critically contest the socio-spatial context and conditions within which conventional Westernised space and architecture emerge.

Possibilities for further research include the exploration of other contemporary pro-poor development practice using the methodology of comparison utilised in this book. This suggests possible engagement with alternative practitioners such as Elemental architecture in Chile, or the work of UTT (Urban Think Tank) in South Africa, in order to contest their spatial practices against aspects of Western spatial theory.

This same examination of contemporary spatial practices is equally able to be directed towards examples drawn from the context of explicitly Westernised space. The opportunity exists for a critical comparison and engagement with alternative, participatory, or grass-roots practices in the UK. Such research could seek to integrate an explicitly critical and reflective platform of collaboration with which to engage with architects, people, and places who are attempting to contest the type of spatial aspirations and themes advocated in this book. Such an engagement with alternative spatial practices might intersect with the work already outlined by Awan et al. (2011), Hyde (2012), Hickey (2012) and others. Yet the textual and comparative reading to critical spatial theory explored in our analysis offers the potential to complement, extend, and challenge the existing academic discourse in this area of spatial agency and practice.

Throughout this book we have introduced examples of alternative spatial practice from Western contexts that each deserve further research. There remains an invaluable opportunity to re-read the history of informal space in the UK from the study of plotlanders by Colin Ward, and both the housing initiatives of Walter Segal and Nabeel Hamdi's work for the LCC in the 1970s. In a more contemporary UK context practices such as Architecture 00, Assemble Studio, practice architecture, etc, require a similar process of critical comparison to spatial theory in order to critically validate their practices. And in an even more practical way, research must be pursued into how and why such practices emerge in order to facilitate the replication and expansion of such exemplary practices. Only in this way can we begin to contest the seemingly endless line of architectural practices defined by the conventions and assumptions of neoliberal economics and capitalist spaces.

The opportunity and necessity exists to question the social, political, and economic contexts in which both alternative practices and individual projects succeed or fail. As a mechanism to bridge the divide between spatial theory and practice it is vital to engage with real-world examples of practice from within Westernised space, and attempt to critically learn from them. Placing such examples in comparison with methodologies of grass-roots and participatory development has the potential to radically improve the potential of such projects in Western space, and help them to achieve socially sustainable change. It also has the potential to question economic assumptions and implications that alternative spatial practice in Western space imposes on those willing to pursue grass-roots and participatory projects and positive spatial agency.

Many such questions remain. What might concepts of dialectical materialism, counter-hegemonic practices, disruptive participation, and textual value imply in the context of Westernised space, social relations, economics, and politics? How will informal spaces and architectures affect the hierarchical planning of the Global North in the near future? And how can we begin to teach our future architectural and spatial practitioners, politicians, and the public about the positive potential of such controversial spatial relations? Questions like these remain outstanding from the outcomes of this book, but are perhaps able to be framed, critiqued, and discussed more positively and pro-actively in the context of the comparisons articulated here.

The observations outlined by the research in this book stand in contradiction to the accepted ideological structures – economic, social, and political – that tend to predominate and prevail in Westernised space and architecture. In the context of Lefebvre, Massey, and wider critical socio-spatial theory, the re-imagination of space and architecture proposed by our study is reliant upon an agency of unknowing, undecideability, and open-ended practices, as exemplified in the works of Hamdi and Turner. Yet challenges to the certainty, cohesion, and authority of the architectural profession as observed in these comparisons offer an inversion to conventional interpretations that are likely to greatly resist change. The challenge therefore remains to confront and contest the social relations of Westernised space; recognising the immense challenge this poses without relinquishing the social agency of architecture to the current state of economic, social, and political neoliberalism. It is hoped that the critical comparison of ideas such as the social agency of small change practices of disruption and the humility of user-choice housing can begin to provide a renewed critical framework for the contestation of Westernised space.

What does this offer as a reflection of the accepted social relations that define Westernised architecture and space? In response to the comparisons and conclusions drawn in this book it is clear that by looking to grass-roots participatory development practices we can begin articulate the positive potential of a political re-imagining of space as a social practice and architecture as a social product.

References

Awan, N., Schneider, T., Till, J., 2011. *Spatial Agency.* Routledge, London.

Featherstone, D., Painter, J., 2013. There is no Point of Departure: The Many Trajectories of Doreen Massey, in: Featherstone, D., Painter, J. (Eds), *Spatial Politics: Essays for Doreen Massey.* Wiley-Blackwell, Chichester.

Goonewardena, K., Kipfer, S., Milgrom, R., Schmid, C., 2008. Globalizing Lefebvre?, in: Goodewardena, K., Kipfer, S., Milgrom, R., Schmid, C. (Eds), *Space, Difference, Everyday Life: Reading Henri Lefebvre.* Routledge, New York.

Hall, S., Massey, D., Rustin, M., 2013. After Neoliberalism: Analysing the Present. *Soundings* 53 (April), 8–22.

Harvey, D., 2005. *A Brief History of Neoliberalism.* Oxford University Press, Oxford.

Hickey, A.A., 2012. *A Guidebook of Alternative Nows*. The Journal of Aesthetics and Protest Press, Los Angeles, CA.

Hyde, R., 2012. *Future Practice: Conversations from the Edge of Architecture*. Routledge, London.

Laclau, E., Mouffe, C., 2001. *Hegemony and Socialist Strategy*. Verso, London.

Lefebvre, H., 1969. *The Explosion*. Monthly Review Press, New York.

Massey, D., 2005. *For Space*. Sage Publications, London.

Spivak, G.C., 1985. Scattered Speculations on the Question of Value. *Diacritics* 15(4), Marx after Derrida, 73–95.

Glossary

To facilitate the simple consumption of this book and its discursive trajectory a series of key terms are here outlined and contextualised. They are included here to allow for an introduction to these ideas whilst maintaining simplicity within the main text. This glossary also reflects an accompanying awareness and acknowledgement of the complexity entailed within such terms. It is an opportunity to recognise that the definitions outlined below, and their use throughout the book, are by no means definitive or universal, nor are they intended to be read in that way.

Development practice

This term recognises a range of socio-spatial practices and methodologies that are utilised to facilitate changes and improvements towards accepted goals of development. Approaches to international development are reflected in the policy priorities of major development organisations such as the UN, World Bank, national and local governments, global NGOs and grass-roots organisations. The most widely acknowledged recent structural identification of development goals are the 'Millennium Development Goals' (MDGs) or the recent parallel incarnation of 'Sustainable Development Goals' (SDGs), which have formed the basis of Western articulations of global development since the millennium summit of the United Nations in 2000. These structural and institutional articulations of development goals traditionally inform the framework for 'on the ground' and front-line actions of development practitioners, and conventional models of development practice in particular.

The action and agency of aid-workers, campaigners, and development practitioners can each be subsumed within the notion of development practice. Thus the spectrum of spatial methodologies and practice is far broader than conventional notions of architectural practice, and provide ways to address the practical and theoretical challenges and complexities of the field of development. By engaging with the diverse identities living in cities of the Global South, development practitioners can broadly be thought of as seeking to generate greater social (and eventual economic) equality and well-being by exploring and facilitating processes of social change, enterprise, and development.

It is in this context that Turner and Hamdi are posited as exemplars of politically and practically alternative spatial agency who each contested the assumptions of development goals. Their approaches are observed and compared as offering an alternative and counter-balance to conventional hierarchical, institutional, and market-led processes of development. Such grass-roots methodologies, observations, and practices can be seen to highlight the social and material reality of rapid urbanisation, diversity, and globalisation. They confront and contend questions of whether economic growth alone is sufficient to address social inequities and promote real sustainable well-being.

Thus the term 'development practices' can also describe non-traditional forms of engagement in social and political spaces of development. As explored throughout this study, they engage in the informal settlements and peripheries of space and culture, whilst also suggesting methodologies that reflect many aspirations of Western spatial theory.

Informal settlements

This term is perhaps most conventionally identified with favelas,[1] barrios,[2] and slums[3] as the most culturally recognisable examples of informally produced settlements. It inherently describes a variety of urban conditions that exist outside the conventions of formal planning, yet the spectrum of informality and difference to formal planning models is far more complex than the structuralist, prescriptive, and negative binary of the terms formal and informal suggests.

Informality is understood by this research through the distinctions of non-traditional and non-hierarchical geometries of power articulated in the creation, occupation, and management of alternative spaces and communities (Baltazar and Kapp, 2007). Thus, in contrast to the methodologies of conventional, economic, and ideologically Westernised planning, the term 'informal settlement' allows various identities to intersect and coalesce around the spatial articulation of socio-economic difference. Subsequently, the idea of informal settlement is utilised in this research in explicit connection with the autonomous and progressive housing models of Turner, and the sustainable community planning of Hamdi.

It is important to note once again that informal settlements should not be understood as merely existing dualistically with formal models of planning. Instead they both exist on a spectrum of legality and illegality, social convention and difference, centre and periphery. Thus it is expressly observed that more formal definitions of informality exist in spatial forms outside professionally, institutionally, and/or commercially based routes of procurement and grounded in individual/community-based self-build (Baltazar and Kapp, 2007, pp. 1–2). Similarly, it is also important to take this opportunity to make clear that the use of and engagement with informal settlements in this book is explicitly not intended to glamorise or romanticise either the idea or reality of life and living conditions faced by millions of people (Baltazar and Kapp, 2007, p. 18).

Global South

This book uses the conventionally accepted terms Global South and Global North to distinguish between the developed first- and second-world economies predominantly found in the North, and the context of the developing third world in the South. While the loose geographical nature of the Global South/North terms is perhaps loaded with political inaccuracy and tension, since the end of the Cold War it has become widely recognised as the most acceptable terminology when discussing global development (Reuveny, 2009). Whilst this distinction is recognised as an overly simplistic socio-economic and political divide, due to the inherently negative implications of the alternative terms 'developed' and 'developing', or 'first-' and 'third-world' economies, North and South have become the most conventionally accepted distinction used in global academic discourse.

The Global North loosely consists of the United States, Canada, Europe, and East Asia,[4] whilst the Global South consists of Africa, Latin America, developing Asia, South America, and the Middle East.[5] The North is generally understood to be formed of richer economies, but also is distinguished by the prevalence of adequate social conditions, food and shelter, and education for populations.[6] The inverse is observed in the Global South, where three-quarters of the world's population control only one-fifth of the world's income. Only 10% of the world's manufacturing industries are both owned and controlled by the South (Therien, 1999).

However, the analysis in this book seeks to frame the use of the terms 'Global North' and 'Global South' through a more progressive academic articulation of the challenges of global capitalism. This articulation would seek to intersect with the discourse of both Mouffe and Massey who interpret the hegemonic characteristics of space being disseminated from nodal points at the heart of geometries of power (Massey, 2004, p. 12; Mouffe, 2013, p. 29). Here it is thus equally important to recognise distinctions between centre and periphery, majority and minority, formal and informal, within the contexts of individual countries, regions, and cities (Ferguson and Gupta, 1992, p. 19). In this articulation it is recognised that elements of the socio-economic and political inequality faced by the Global South are recognised within the borders of the Global North territories.[7]

Spatial practice

Similar to informal settlements, spatial practice is a term utilised in this book to cover a variety of alternative practical engagements with questions of space and the built environment. Thus, subsumed under this term are practices explored in the context of both the Global North and the Global South. From a perspective of Westernised space the positive potential of social agency and spatial practice has already been eloquently articulated by Nishat Awan et al. (2011), and continues to be explored theoretically in the works of

Rory Hyde (2012) and Amber Hickey (2012), among others. The notion of social agency provides an approachable concept with which to interpret our comparisons with development practice. The positive social agency that Awan et al. articulated is a conception of alternative spatial practices that includes many examples drawn from development practice that connect with the trajectory of our discourse.

Thus the spirit and agency of alternative spatial practices can be observed within the development practices of Turner and Hamdi explored in this book. Here, the social and economic improvement of space is understood to transcend architecture as the production of conventional built form. Instead, such spatial practice seeks to produce and practice change through a spatial agency that does not rely on conventional models of economic or social value, and which contest the social, political, and economic contexts in practices of the everyday that are grounded in concrete reality.

Western / Westernised

The use of the term Western in this book is equally as complicated as the distinctions made above concerning Global North and Global South. In general the term 'Westernised' is used to denote the conventionally accepted social, political, and economic spaces, practices, and institutions that have accompanied the advent of neoliberal capitalism (Harvey, 2005; Ronneberger, 2008). Within this articulation is a recognition that Western space, values, and ideals have been readily adopted throughout the world, becoming nodal points of money, power, and homogenisation that can be seen equally in London, New York, and Beijing, as they can in Lagos, Caracas, and Mexico City.

Conversely however, it must also be recognised that there are elements of difference and alterity – quite often exemplified in spatial practices – that exist in contradiction to the neoliberal model of Westernised space. In the Global North these elements can be observed as spatial tactics working within the confines of neoliberal strategies (de Certeau, 1988, p. 29), whereas in the Global South the balance and inequity of neoliberal space is highlighted in spatial points of far more concentrated and explicit dominance and inequality. Examples of this include the proliferation of skyscrapers as symbols of economic vitality and development, perhaps most notable in the contradictions between favelas and oligarchic residential towers for example in Dharavi in Mumbai, India. Examples can also be drawn from the intense inequality of new mega-cities like Lagos, Nigeria as well as less explicitly successful attempts at neoliberalism such as the Torre David in Caracas (Brillembourg and Klumpner, 2012). In general however, the use of Western or Westernised in this book is intended to convey the unquestioned sense of conventional inevitability that Massey describes as accompanying the advent of globalisation at the expense of the positive potential and political necessity of multiplicity (2005, p. 4).

Notes

1 Favela is a Portuguese term for urban slum conditions in Brazil. The first noted favelas were built by soldiers returning from the war of Canudos, who, finding they had nowhere to live, built temporary dwellings upon Providence Hill in Rio de Janeiro, which was noted for having many favela trees upon it.

2 Barrio was originally a Portuguese term for a city community or region. However the increasingly negative identity of barrios in comparison to Western ideas of regions emerges from derogatory identification of early informal settlements as *barrios Africanos* (African neighbourhoods).

3 The term slum is thought to have originally meant room, which later evolved to 'back slum' with the meaning of 'back alley, for street people'. See: Slum. *Etymology Dictionary*, Douglas Harper (2001).

4 The economic and political implications of this identity can be observed in the Global North pertaining to almost all of the permanent members of the UN security council, and all members of the G8.

5 Further note might be taken of the increasing importance of the emerging economic power of the so-called BRIC nations: Brazil, Russia, India, and China. This distinction is primarily of economic importance in terms of global manufacturing and does not reflect the questions of poverty that pervade such countries. As such this distinction remains somewhat unhelpful in the broad discussion of issues of global inequality and development within this book.

6 It is observed that 95% of the Global North adheres to international standards in these issues, whereas the Global South is widely recognised as only achieving those standards for approximately 5% of its population (Oluwafemi, 2012, p. 47).

7 For recent references to this issue, see McElwee (2014). However, similar critique could be brought against the political and legal situations in Russia (notably the contemporary issues concerning the Socchi 2014 Winter Olympics, the Ukraine and Crimea, and the various arrests of political antagonists such as Mikhail Khodorkovsky and the lesbian punk rock band Pussy Riot) and China (the forced evictions to make way for the 2008 Beijing Olympic Park, and various continued economic challenges coupled with widespread control of political state media etc.).

References

Awan, N., Schneider, T., Till, J., 2011. *Spatial Agency*. Routledge, London.

Baltazar, A.P., Kapp, S., 2007. Learning from 'Favelas': The Poetics of Users' Autonomous Production of Space and the Non-ethics of Architectural Interventions. In Proceedings of the International Conference Reconciling Poetics and Ethics in Architecture (McGill University, Canada, September 2007). Available at http://www. arch.mcgill.ca/theory/conference/papers.htm (accessed 1 March 2016).

Brillembourg, A., Klumpner, H. (Eds), 2012. *Torre David: Anarcho Vertical Communities*. Lars Müller, Zurich.

de Certeau, M., 1988. *The Practice of Everyday Life*. University of California Press, Berkeley.

Ferguson, J., Gupta, A., 1992. Beyond 'Culture': Space, Identity and the Politics of Difference. *Cultural Anthropology* 7, 6–23.

Harvey, D., 2005. *A Brief History of Neoliberalism*. Oxford University Press, Oxford.

Hickey, A.A., 2012. *A Guidebook of Alternative Nows*. The Journal of Aesthetics and Protest Press, Los Angeles, CA.

Hyde, R., 2012. *Future Practice: Conversations from the Edge of Architecture*. Routledge, London.

Massey, D., 2004. Geographies of Responsibility. *Geografiska Annaler Series B: Human Geography* 86, 5–18.

Massey, D., 2005. *For Space.* Sage Publications, London.

McElwee, S., 2014. Six Ways America Is Like a Third-World Country. *Rolling Stone*, 5 March. Available at http://www.rollingstone.com/politics/news/six-ways-america-is-like-a-third-world-country-20140305 (accessed 15 March 2016).

Mouffe, C., 2013. Space, Hegemony and Radical Critique, in: Featherstone, D., Painter, J. (Eds), *Spatial Politics: Essays for Doreen Massey.* Wiley-Blackwell, Chichester.

Oluwafemi, M., 2012. *Globalization: The Politics of Global Economic Relations and International Business.* Carolina Academic, Durham, NC.

Reuveny, R.X., 2009. The North–South Divide and International Studies: A Symposium. *International Studies Review* 9(4), 556–564.

Ronneberger, K., 2008. Henri Lefebvre and Urban Everyday Life: In Search of the Possible, in: Goodewardena, K., Kipfer, S., Milgrom, R., Schmid, C. (Eds), Kipfer, S., Brenner, N. (Trans.), *Space, Difference, Everyday Life: Reading Henri Lefebvre.* Routledge, New York.

Therien, J.-P., 1999. Beyond the North–South Divide: The Two Tales of World Poverty. *Third World Quarterly* 20(4), 723–742.

Contextualisation of key protagonists

The research for this book revealed a rich constellation of connections, similarities, and intersections between its four main protagonists, yet these relationships are situated within a wider reading of surrounding discourse. This wider context provides a foundation for the points of focused comparison and analysis between theory and practice. It was necessary and valuable to explore such a broad theoretical context in order to support the main thrust of the book and its specific focus on the relationships between the four key protagonists.

For example, the valuable text *Spatial Agency* by Nishat Awan et al. (2011) provides a useful frame of reference when introducing the concept of alternative spatial practice in Westernised space and spatial theory. The premise of *Spatial Agency* was to provide a broad and explorative compendium of similarly framed alternative spatial practices, and the timing, success, and value of this text can be linked precisely to its broad narrative. Yet *Spatial Agency* was never intended to provide focused in-depth scrutiny of the theoretical connections and themes that emerge from close study of such examples. In contrast, the methodology of our book is explicitly intended to provide such an in-depth exploration. Thus, in revealing new connections between specific trajectories of theoretical and practical spatial discourse our comparative analysis provides a valuable addition to the existing literature surrounding alternative spatial agency and practice.

The wide literature review supporting this book similarly observes connections with other theoretical discourses on space. For example, this research offers opportunities for detailed exploration of the work of David Harvey, which can be considered as an intermediary between the discourses of Lefebvre and Massey. From his early discussions of *Social Justice and the City* (2010), through to his more explicit contemporary writings such as *Rebel Cities* (2012), Harvey's work inevitably interconnects with the intentions of our study. However, the utilisation of Massey as a primary protagonist instead of Harvey reflects both the emergent nature of the evolution of this book, and also an observation that Harvey does not often provide the same positive perspective and analysis of space that this research observed and valued in Massey's discourse.

Wider connections can also be made to the work of Kim Dovey whose discourse is recognised as a valuable contemporary contribution to the politics of urban space (2012; Dovey and Sandercock, 2002). It is compelling that Dovey's work is often similarly engaged in discussion of urban informality and alternative models of urban form (2012; Dovey and King, 2013). These recent writings have provided valuable complementary reading in the contemporary contextualisation within our study. Whilst Dovey's texts are not explored in our study in any explicit detail, they remain valuable points of support as part of the wider context of our research intentions and aspirations.

Building upon this brief review of just a sample of the wider discourse surrounding the research in this book, what follows are similarly brief examinations of the four main protagonists, and an assessment of existing primary and secondary literature that intersects with the connections examined in the course of our study. Whilst throughout the research there have been numerous points where the interdisciplinary connections framed in our book are tangible, the exact critical connections and comparisons raised in this study have not been observed elsewhere in the literature review and thus, it is hoped, will offer new contributions to existing discourse, and hopefully reveal the value of an emerging discourse and international comparison of the spaces and spatial relations that define our cities, both formal and informal.

John F.C. Turner

Turner is widely recognised as a key protagonist in the development of alternative and socially progressive housing models in Latin America in the 1960s (Sanyal et al., 2008, p. 16). His extensive writing on housing and community organisation was influenced by his experiences working in the squatter settlements of Peru from 1957 to 1965. As both Ray Bromley (2003) and Richard Harris (2003, 1999, 1998) note, Turner's work must be contextualised against an understanding of Peru as a world leader in housing policy, community development, and self-help in the 1950s and 1960s, as well as observing the influence of Peruvian architects and urban theorists Pedro Beltrán, Carlos Delgado, and Fernando Belaúnde.[1]

Turner widely acknowledges his theoretical debt to the works of Lewis Mumford (1938) and Patrick Geddes (1949) as well as more subtle references to the anarchist works of Peter Kropotkin (2006), Ivan Illich (1976), and Giancarlo de Carlo (1949). Yet Turner's work also owes a great theoretical debt to the sociological works of William Mangin (1967) whose study of the evolution of housing in Latin America would become a vital theoretical basis for Turner's later analysis.

Whilst only a limited number of primary sources from Turner exist they are exemplary in forming a foundational premise of the political and economic logic of his approach to space (1976, 1972, 1963). His work and discourse in the 1960s and 1970s was notably reflected on and supported by Colin Ward

(1972), whose work from the same period sought to articulate a conception and positive contestation of anarchist housing as a proposition for the UK. Ward himself was an influential academic protagonist in the discourse of post-Second World War housing in the UK (1976), key practical realisations of which can be read in a small number of seminal participatory architecture projects in 1960s UK. These are exemplified in the work by Ralph Erskine at the Byker Wall housing project in Newcastle (1968), Cedric Price's speculative projects of the Potteries Think-belt (1969) and Fun Factory (1961), and Nabeel Hamdi's work for the GLC in the 1970s and 1980s, including the Adelaide Road Housing programme under the PSSHAK system (Primary Support Structures and Housing Assembly Kits) – a practical interpretation of John Habraken's theories of support and infill (Hamdi, 1991).

At the peak of his professional and academic popularity in the 1970s Turner's discourse was also subject to a variety of criticisms (Harms, 1982, 1976; Ward, 1982), most notably by the avowed neo-Marxist Rod Burgess (1982, 1978a, 1978b). This critique is explored further in chapter one yet it is important here to note the complex historical critical context in which this book frames Turner's literature.

In more contemporary discourse Turner's work is re-emerging as a renewed source of both professional and academic interest as the positive and negative issues of informal architecture are observed as becoming increasingly prevalent. The more connected world of instantaneous images and media has confronted Western audiences with the global inequality of divisions of labour and living conditions (Davis, 2007; Neuwirth, 2012, 2006; Pugh, 2000; Roy, 2011, 2005).

As such, the implications of Turner's work have been reviewed both practically and theoretically through various contributions (Baltazar et al., 2008; Baltazar and Kapp, 2007; Fernández-Maldonado, 2007; Hodkinson, 2012; Lyons et al., 2010; Ward, 2008). Yet the research in this book observed that even with this renewed interest, the disjunction between analysis of the theoretical and practical implications of Turner's work remains largely unchanged and constrained by disciplinary boundaries. The new interdisciplinary comparisons in our study thus contribute to and challenge this existing discourse. It explicitly engages with and contests the inherent assumptions and disjunctions found in the gap between spatial theory and practice. In doing so it tries to re-imagine and re-value the theoretical implications of Turner's work.

Henri Lefebvre

Henri Lefebvre's work has defined him as one of the pre-eminent French Marxist philosophers and sociologists of the twentieth century, and he is best known for pioneering critiques of everyday life, rights to the city, and the social production of space. His work was most notably the subject of great academic interest in the Anglophone world after the 1991 publication of the first English translation of *The Production of Space* (1991). Yet the true scope, scale, and complexity of Lefebvre's interrogations of space have only begun to be critically understood more broadly following the various examinations made by Stuart Elden (2004), Neil Brenner (1997; Brenner and Elden, 2009; Ronneberger, 2008), Rob Shields (1999), Andy Merrifield (2006), Lukasz Stanek (2014), and Kanishka Goonewardena et al. (2008). The research trajectory of this book sought to utilise these texts in connection with a variety of Lefebvre's original source materials in order to provide a robust foundation for the interdisciplinary comparisons and connections posited.

However, it is important to state that the interdisciplinary intention of this research is explicitly not intended as a means to critique the work of Lefebvre. As such the choice of source material drawn from Lefebvre has been targeted in order to frame the comparisons rather than to provide a complete analysis of his entire discourse. This has meant a rather unconventional engagement with some of Lefebvre's less prominent texts, including his early work *Dialectical Materialism* (1968), his critical extension of Marxism in *The Survival of Capitalism* (1976), as well as his more prominently observed works on the city and space (2003, 1996, 1991, 1969).

The interdisciplinary and explorative focus of this book has also meant that Lefebvre's work cannot be explored here in its entirety. The most notable implication of this has been the only limited and implicit connections made towards his discourse concerning everyday life (2004, 2002). However, the various themes of festival, spontaneity, and everyday life that occur throughout *The Production of Space*, as well as broad references from secondary resources, have allowed implicit moments of utilisation of such themes in this book. Here it is hoped that the combination of a robust analysis of explicit primary sources and the broader contextualisation of secondary sources have provided a viable foundation for the interdisciplinary comparisons drawn in this research.

Whilst the comparative connections from Lefebvre to Turner explored in this book remain a novel inquiry from the perspective of both spatial theory and development, this comparison is bolstered by the recent prominent conceptualisation of Lefebvre in Andrea Cornwall's analysis of the 'invited spaces' of participatory development (2004). Whilst this remains a markedly singular connection observed by this research it provides a sense of the opportunity that the interdisciplinary comparison of our study offers to the usually overly theoretical and practical discussions of space.

In a similar way, the links between Massey and Lefebvre are surprisingly somewhat tangential, especially given the theoretical intersections of their respective discourses on space outlined in this book. Whilst references to Lefebvre do appear in the work of Massey and secondary discussions of her, they are remarkably isolated and minimal (Massey, 2005, p. 17, 1999, pp. 2, 3, 6). Massey's own articulations of Marxism and spatial relations appear to bypass Lefebvre, instead being a product of her extensive study of Marxism itself, and subsequently the works of Laclau and Mouffe (2001, pp. 90, 109) which are themselves reworkings of Louis Althusser's and Antonio Gramsci's Marxist re-contextualisation.

This observation perhaps provides a rationale for the otherwise glaring disconnection between Massey and Lefebvre (Featherstone and Painter, 2013, p. 4; Saldanha, 2013, p. 48). Yet Massey is known to be overtly aware of the works of Lefebvre and seems to have built certain aspects of her interpretations of space on Lefebvre's advocacy of space as emergent and real, with Massey adding a sense of density in the unfolding of its multiplicity (Grossberg, 2013, p. 34). There are also further overt references in Massey's work to Lefebvre's post-structural considerations in *Beyond Structuralism* (Lefebvre, 2006, p. 38), yet the lack of critical comparisons between them remains conspicuous. Given the interconnected comparison proposed in our research and the clear intersection of their respective discourses on conceptions of the positive potential of space, the analysis of this book seeks to begin to confront and rectify this gap in contemporary spatial discourse.

Doreen Massey

Massey's writings on social science, feminism, and post-colonial and Marxist geography emerged prominently in the 1980s with her work the *Spatial Divisions of Labour* (1984). This ground-breaking examination explored the geographical implications of regional inequality in the aftermath of the post-industrial restructuring of the UK in the 1970s. It is here that Massey began to articulate the concept of power-geometry as informing patterns of unequal relationships from the perspective of a Marxist political economy (Massey, 1999; Saldanha, 2013, p. 48).

From these beginnings, Massey's discourse has become increasingly rich, provocative, and multidimensional, first noted through her engagements with gender in the text *Space, Place and Gender* (1994), and the globalised dialogues of *For Space* and *World City* (2005, 2007). The variety of themes and interdisciplinary connections explored in these creative texts by Massey provide a relatively complex constellation of ideas and issues with which our book has attempted to traverse. It is in this context that the very recent publication of *Spatial Politics: Essays for Doreen Massey* (2013) has been a most welcome multidisciplinary reflection on the connections and impact of Massey's work. This text provides perhaps the first grounding analysis and discussion of Massey and is warmly welcomed as it helpfully reinforces many of the links suggested by the research in our study.

Reading Massey's work in the context of the methodology of critical comparison explored in this book has allowed significant focus to be given to the text *For Space*, specifically because it provides a framework of analysis and references from which to draw connections to and from her discussions of space. This complexity is reflected in the title of Featherstone and Painter's introduction to their *Spatial Politics*, 'There is no Point of Departure: The Many Trajectories of Doreen Massey' (2013). Here the sheer variety and richness of Massey's numerous articles, collaborations, and interconnections are recognised as a reflection of the interdisciplinary innovation that Massey has brought to radical geography. In this context, *Architecture and Space Re-imagined* is explicitly not an attempt to engage in a critical examination of the breadth of Massey's discourse. Instead it is an opportunity to pursue a trajectory of what Arturo Escobar describes as 'the emergent ways of talking about relationality' (2013, p. 170) that have prospered in the wake of Massey's discourse.

Perhaps the most interesting and explicit connection of Massey to development practice comes from her most recent work and engagements in the Global South. Since 2007, Massey's discourse concerning the global politics of inequality has been explicitly explored in her work in Venezuela, where her concept of power-geometry has been utilised in Hugo Chaves'[2] forming of the fifth republic movement (2011). In Venezuela and increasingly across socialist governments of the Global South it is widely observed that Massey's theories have been influential as a means of thinking about and engaging with programmes of decentralisation and equalisation of political power.[3]

Nabeel Hamdi

Hamdi is perhaps the least academically discussed protagonist of this book. His key publications can be counted on one hand and yet his influence in the teaching and dissemination of development as a spatial practice has been profound. This is most notably observed through his immense contributions as a pedagogue on the subject of development practice at Oxford Brookes in 1992 and later at the Development Planning Unit at London UCL, as well as now being a pre-eminent visiting lecturer and speaker on development.

His first published text was the influential *Housing Without Houses* (1991), which provides an almost unmatched technical analysis of global self-built housing as a universal human exercise.[4] This text marked a timely reflection upon the loss of social and political engagement that Hamdi appears to have encountered and challenged during his time working with the Greater London Council (GLC) on flexible and participatory housing during the 1970s and early 1980s (Wainright, 2013). This text was followed by a broader analysis of planning, cities, and community with long-time collaborator Reinhard Goethert, evolving from early papers published in *Habitat International* into the later broad and provocative text, *Action Planning for Cities* (1986, 1989, 1997).

These examinations provided the foundation for his later, more widely observed texts, namely *Small Change* (2004) and *The Placemaker's Guide to Building Community* (2010). What is notable throughout all of Hamdi's published work is the explicitly practical nature of the discourse, which utilises his own experiences, alongside the voices of others working with him, to describe the positive potential of alternative spatial practices of development. It is from these practical thematic studies and analyses that the comparative threads of our study are drawn, utilising Hamdi's self-reflective analysis not only of the places of development but the process of listening, learning, and engaging in social practices of partnership in the course of pursuing socially sustainable enterprise and development. Hamdi's new book *The Spacemaker's Guide to Big Change: Design and Improvisation in Development Practice* (2014), was released after the completion of the research for this book, yet a brief review of this new text only stands to reinforce the comparisons explored in our book, and even offers new opportunities for future research and analysis.

Notes

1 Fernando Belaúnde trained as an architect in the USA in the 1930s and is notable for becoming president of Peru first from 1963 to 1968 before being deposed by a military coup. He was then later re-elected in 1980 after eleven years of military rule, serving till 1985. Widely recognised for his personal integrity and his commitment to the democratic process, he formed the moderate right central political party Acción Popular in 1956 as a reformist alternative to the status quo conservative forces and the populist American Popular Revolutionary Alliance party.
2 Chaves was a known Marxist and his popular Chavista revolution forms part of the contemporary 'pink-tide' of left-wing and socialist democratic movements at work in Latin and Southern America.
3 The fourth of the 'five motors of revolution' was defined as 'The new power-geometry: the socialist re-organisation of the national political geography' (Menendez, 2013).
4 Surpassing the more visual work of Rudolfsky (1987) which, whilst being read more widely, is in comparison a less rigorous technical examination than that offered by Hamdi.

References

Awan, N., Schneider, T., Till, J., 2011. *Spatial Agency*. Routledge, London.
Baltazar, A.P., Kapp, S., 2007. Learning from 'Favelas': The Poetics of Users' Autonomous Production of Space and the Non-ethics of Architectural Interventions. In Proceedings of the International Conference Reconciling Poetics and Ethics in Architecture (McGill University, Canada, September 2007). Available at http://www.arch.mcgill.ca/theory/conference/papers.htm (accessed 1 March 2016).
Baltazar, A.P., Kapp, S., Morado, D., 2008. Architecture as Critical Exercise: Little Pointers Towards Alternative Practices. *Field* 2, 7–30.
Brenner, N., 1997. Global, Fragmented, Hierarchical: Henri Lefebvre's Geographies of Globalisation. *Public Culture* 10, 135–167.
Brenner, N., Elden, S., 2009. Introduction, in: *Henri Lefebvre – State, Space, World: Selected Essays*. University of Minnesota Press, Minneapolis.
Bromley, R., 2003. Peru 1957–1977: How Time and Place Influenced John Turner's Ideas on Housing Policy. *Habitat International* 27, 271–292.
Burgess, R., 1978a. Petty Commodity Housing or Dweller Control. *World Development* 6, 1105–1133.
Burgess, R., 1978b. Self-Help Housing. A New Imperialist Strategy? A Critique of the Turner School. *Antipode* 9, 50–60.
Burgess, R., 1982. Self-help Housing Advocacy: A Curious Form of Radicalism. A Critique of the Work of John F.C. Turner, in: Ward, P.M. (Ed.), *Self-Help Housing: A Critique*. Mansell, London.

Cornwall, A., 2004. Spaces for Transformation? Reflections on Issues of Power and Difference in Participation in Development, in: Hickey, S., Mohan, G. (Eds), *Participation: From Tyranny to Transformation*. Zed Books, London.

Davis, M., 2007. *Planet of Slums*, Reprinted Edition. Verso, London.

de Carlo, G., 1949. The Housing Problem in Italy. *Freedom*, 12 June, 19 June.

Dovey, K., 2012. *The Temporary City*. Routledge, London.

Dovey, K., King, R., 2013. Interstitial Metamorphoses: Informal Urbanism and the Tourist Gaze. *Environment and Planning D: Society and Space* 31, 1022–1040.

Dovey, K., Sandercock, L., 2002. Hype and Hope. City: Analysis of Urban Trends, Culture, Theory, *Policy, Action* 6, 83–101.

Elden, S., 2004. *Understanding Henri Lefebvre: Theory and the Possible*. Continuum, London.

Featherstone, D., Painter, J., 2013. There is no Point of Departure: The Many Trajectories of Doreen Massey, in: Featherstone, D., Painter, J. (Eds), *Spatial Politics: Essays for Doreen Massey*. Wiley-Blackwell, Chichester.

Fernández-Maldonado, A.M., 2007. Fifty Years of Barriadas in Lima: Revisiting Turner and De Soto. Paper presented at the ENHR 2007 International Conference on Sustainable Urban Areas. Rotterdam, the Netherlands.

Geddes, P., 1949. *Cities in Evolution*, 2nd Edition. Williams and Norgate, London.

Goonewardena, K., Kipfer, S., Milgrom, R., Schmid, C. (Eds), 2008. *Space, Difference, Everyday Life: Reading Henri Lefebvre*. Routledge, New York.

Grossberg, L., 2013. Theorising Context, in: Featherstone, D., Painter, J. (Eds), *Spatial Politics: Essays for Doreen Massey*. Wiley-Blackwell, Chichester.

Hamdi, N., 1991. *Housing Without Houses: Participation, Flexibility, Enablement*. Van Nostrand Reinhold, New York.

Hamdi, N., 2004. *Small Change*. Earthscan, London.

Hamdi, N., 2010. *The Placemaker's Guide to Building Community*. Earthscan, London.

Hamdi, N., 2014. *The Spacemaker's Guide to Big Change: Design and Improvisation in Development Practice*. Earthscan tools for community planning. Routledge, New York.

Hamdi, N., Goethert, R., 1986. Implementation: Theories, Strategies and Practice. *Habitat International* 9, 33–44.

Hamdi, N., Goethert, R., 1989. The Support Paradigm for Housing and its Impact on Practice: The Case in Sri Lanka. *Habitat International* 13, 19–28.

Hamdi, N., Goethert, R., 1997. *Action Planning for Cities: A Guide to Community Practice*. John Wiley, Chichester.

Harms, H., 1976. Limitations of Self-Help. *Architectural Design* 46, 230–231.

Harms, H., 1982. Historical Perspectives on the Practice and Politics of Self-Help Housing, in: Ward, P.M. (Ed.), *Self-Help Housing: A Critique*. Mansell, London.

Harris, R., 1998. The Silence of the Experts: 'Aided Self-Help Housing' 1939–1954. *Habitat International* 22, 165–189.

Harris, R., 1999. Slipping Through the Cracks: The Origin of Aided Self-Help Housing 1918–1953. *Housing Studies* 14, 281–309.

Harris, R., 2003. A Double Irony: The Originality and Influence of John F.C. Turner. *Habitat International* 27, 245–269.

Harvey, D., 2010. *Social Justice and the City*, Revised Edition. University of Georgia Press, Athens, GA.

Harvey, D., 2012. *Rebel Cities*. Verso, London.

Hodkinson, S., 2012. The Return of the Housing Question. *Ephemera: Theory and Politics in Organization* 12, 423–444.

Illich, I., 1976. *Deschooling Society*. The Philips Park Press, Manchester, UK.

Kropotkin, P., 2006. *Mutual Aid: A Factor of Evolution*. Dover Books on History, Political and Social Science, Dover, New York.

Laclau, E., Mouffe, C., 2001. *Hegemony and Socialist Strategy*. Verso, London.

Lefebvre, H., 1968. *Dialectical Materialism*. Jonathon Cape, London.

Lefebvre, H., 1969. *The Explosion*. Monthly Review Press, New York.

Lefebvre, H., 1976. *The Survival of Capitalism*. Allison and Busby, London.

Lefebvre, H., 1991. *The Production of Space*. Blackwell Publishing, Oxford.

Lefebvre, H., 1996. *Writings on Cities*. Blackwell Publishing, Oxford.

Lefebvre, H., 2002. *Critique of Everyday Life*. Verso, London.

Lefebvre, H., 2003. *The Urban Revolution*. University of Minnesota Press, Minneapolis.

Lefebvre, H., 2004. *Rythymanalysis*. Continuum, London.

Lefebvre, H., 2006. *Key Writings*, 3rd Edition. Athlone Contemporary European Thinkers, Continuum, London.

Lyons, M., Schilderman, T., Boano, C. (Eds), 2010. *Building Back Better*. South Bank University: Practical Action Publishing, London.

Mangin, W., 1967. Latin American Squatter Settlements: A Problem and a Solution. *Latin American Research Review* 2, 65–98.

Massey, D., 1984. *Spatial Divisions of Labor: Social Structures and the Geography of Production*. Methuen, Basingstoke.

Massey, D., 1994. *Space, Place and Gender*. Polity Press, Cambridge.

Massey, D., 1999. *Power-Geometries and the Politics of Space-Time*. Department of Geography, University of Heidelberg, Heidelberg.

Massey, D., 2004. Geographies of Responsibility. *Geografiska Annaler Series B: Human Geography* 86, 5–18.

Massey, D., 2005. *For Space*. Sage Publications, London.

Massey, D., 2007. *World City*. Polity Press, Cambridge.

Massey, D., 2011. A Counterhegemonic Relationality of Place, in: McCann, E., Ward, K. (Eds), *Mobile Urbanism: Cities and Policymaking in the Global Age*. Globalization & Community Series, University of Minnesota Press, Minneapolis.

Menendez, R., 2013. The Social Transformation of Venezuela: The Geographical Dimension of Political Strategy, in: Featherstone, D., Painter, J. (Eds), *Spatial Politics: Essays for Doreen Massey*. Wiley-Blackwell, Chichester.

Merrifield, A., 2006. *Henri Lefebvre – A Critical Introduction*. Routledge, New York.

Mumford, L., 1938. *The Culture of Cities*. Martin Secker, London.

Neuwirth, R., 2006. *Shadow Cities: A Billion Squatters, A New Urban World*, New Edition. Routledge, New York.

Neuwirth, R., 2012. *Stealth of Nations: The Global Rise of the Informal Economy*, Reprinted Edition. Anchor Books, New York.

Pugh, C., 2000. Squatter Settlements: Their Sustainability, Architectural Contributions, and Socio-economic Roles. *Cities* 17, 325–337.

Ronneberger, K., 2008. Henri Lefebvre and Urban Everyday Life: In Search of the Possible, in: Goonewardena, K., Kipfer, S., Milgrom, R., Schmid, C. (Eds), Kipfer, S., Brenner, N. (Trans.), *Space, Difference, Everyday Life: Reading Henri Lefebvre*. Routledge, New York.

Roy, A., 2005. Urban Informality: Toward an Epistemology of Planning. *Journal of the American Planning Association* 71, 147–158.

Roy, A., 2011. Slumdog Cities: Rethinking Subaltern Utopianism. *International Journal of Urban Regional Research* 35, 223–238.

Rudolfsky, B., 1987. *Architecture Without Architects: A Short Introduction to Non-Pedigreed Architecture*, Reprinted Edition. University of New Mexico Press, Albuquerque.

Saldanha, A., 2013. Power Geometry as Philosophy of Space, in: Featherstone, D., Painter, J. (Eds), *Spatial Politics: Essays for Doreen Massey.* Wiley-Blackwell, Chichester.

Sanyal, B., Rosan, C., Vale, L.J. (Eds), 2008. *Planning Ideas That Matter: Livability, Territoriality, Governance and Reflective Practice.* MIT Press, Cambridge, MA.

Shields, R., 1999. *Lefebvre, Love & Struggle: Spatial Dialectics.* Routledge, London.

Stanek, Ł.,2014. *Urban Revolution Now: Henri Lefebvre in Social Research and Architecture.* Ashgate, Farnham, UK.

Turner, J.F.C., 1963. Dwelling Resources in South America. *Architectural Design* 8, 360–393.

Turner, J.F.C., 1972. The Re-Education of a Professional, in: Fichter, R., Turner, J.F. (Eds), *Freedom To Build.* Macmillan Education, New York.

Turner, J.F.C., 1976. *Housing by People: Towards Autonomy in Building Environments.* Marion Boyars, London.

Wainright, H., 2013. Place Beyond Place and the Politics of 'Empowerment', in: Featherstone, D., Painter, J. (Eds), *Spatial Politics: Essays for Doreen Massey.* Wiley-Blackwell, Chichester.

Ward, C., 1972. Preface, in: Fichter, R., Turner, J.F. (Eds), *Freedom To Build.* Macmillan Education, New York.

Ward, C., 1976. *Housing: An Anarchist Approach.* Freedom Press, London.

Ward, P.M. (Ed.), 1982. *Self-Help Housing: A Critique.* Mansell, London.

Ward, P.M., 2008. Self-Help Housing Ideas and Practice in the Americas, in: Sanyal, B., Vale, L.J., Rosan, C. (Eds), *Planning Ideas that Matter: Livability, Territoriality, Governance and Reflective Practice.* MIT Press, Cambridge, MA.

Index